Tucholsky   Wagner   Zola   Scott   Sydow   Freud   Schlegel
Turgenev   Wallace   Fonatne
Twain   Walther von der Vogelweide   Fouqué   Friedrich II. von Preußen
Weber   Freiligrath   Frey
Fechner   Fichte   Weiße Rose   von Fallersleben   Kant   Ernst   Richthofen   Frommel
Hölderlin
Engels   Fielding   Eichendorff   Tacitus   Dumas
Fehrs   Faber   Flaubert   Eliasberg   Ebner Eschenbach
Feuerbach   Maximilian I. von Habsburg   Fock   Eliot   Zweig   Vergil
Goethe   Ewald   London
Mendelssohn   Balzac   Shakespeare   Elisabeth von Österreich   Dostojewski   Ganghofer
Trackl   Lichtenberg   Rathenau   Doyle   Gjellerup
Mommsen   Stevenson   Tolstoi   Lenz   Hambruch   Hanrieder   Droste-Hülshoff
Thoma   von Arnim
Dach   Verne   Hägele   Hauff   Humboldt
Karrillon   Reuter   Rousseau   Hagen   Hauptmann   Gautier
Garschin   Defoe   Hebbel   Baudelaire
Damaschke   Descartes   Hegel   Kussmaul   Herder
Wolfram von Eschenbach   Darwin   Dickens   Schopenhauer   Rilke   George
Bronner   Melville   Grimm   Jerome   Bebel   Proust
Campe   Horváth   Aristoteles   Voltaire   Federer
Bismarck   Vigny   Barlach   Heine   Herodot
Gengenbach   Tersteegen   Grillparzer   Georgy
Storm   Casanova   Lessing   Langbein   Gilm   Gryphius
Chamberlain   Lafontaine
Brentano   Claudius   Schiller   Kralik   Iffland   Sokrates
Strachwitz   Bellamy   Schilling
Katharina II. von Rußland   Gerstäcker   Raabe   Gibbon   Tschechow
Löns   Hesse   Hoffmann   Gogol   Wilde   Gleim   Vulpius
Luther   Heym   Hofmannsthal   Klee   Hölty   Morgenstern
Roth   Heyse   Klopstock   Kleist   Goedicke
Luxemburg   Puschkin   Homer   Mörike   Musil
Machiavelli   La Roche   Horaz
Navarra   Aurel   Musset   Kierkegaard   Kraft   Kraus
Nestroy   Marie de France   Lamprecht   Kind   Kirchhoff   Hugo   Moltke
Nietzsche   Nansen   Laotse   Ipsen   Liebknecht
Marx   Lassalle   Gorki   Klett   Leibniz   Ringelnatz
von Ossietzky   May   vom Stein   Lawrence   Irving
Petalozzi   Platon   Pückler   Michelangelo   Knigge   Kafka
Sachs   Poe   Liebermann   Kock   Korolenko
de Sade   Praetorius   Mistral   Zetkin

The publishing house tredition has created the series **TREDITION CLASSICS**. It contains classical literature works from over two thousand years. Most of these titles have been out of print and off the bookstore shelves for decades.

The book series is intended to preserve the cultural legacy and to promote the timeless works of classical literature. As a reader of a **TREDITION CLASSICS** book, the reader supports the mission to save many of the amazing works of world literature from oblivion.

The symbol of **TREDITION CLASSICS** is Johannes Gutenberg (1400 – 1468), the inventor of movable type printing.

With the series, tredition intends to make thousands of international literature classics available in printed format again – worldwide.

All books are available at book retailers worldwide in paperback and in hardcover. For more information please visit: www.tredition.com

tredition was established in 2006 by Sandra Latusseck and Soenke Schulz. Based in Hamburg, Germany, tredition offers publishing solutions to authors and publishing houses, combined with worldwide distribution of printed and digital book content. tredition is uniquely positioned to enable authors and publishing houses to create books on their own terms and without conventional manufacturing risks.

For more information please visit: www.tredition.com

French Polishing and Enamelling
A Practical Work of Instruction

Richard Bitmead

Imprint

This book is part of the TREDITION CLASSICS series.

Author: Richard Bitmead
Cover design: toepferschumann, Berlin (Germany)

Publisher: tradition GmbH, Hamburg (Germany)
ISBN: 978-3-8491-4988-8

www.tredition.com
www.tredition.de

Copyright:
The content of this book is sourced from the public domain.

The intention of the TREDITION CLASSICS series is to make world literature in the public domain available in printed format. Literary enthusiasts and organizations worldwide have scanned and digitally edited the original texts. tredition has subsequently formatted and redesigned the content into a modern reading layout. Therefore, we cannot guarantee the exact reproduction of the original format of a particular historic edition. Please also note that no modifications have been made to the spelling, therefore it may differ from the orthography used today.

AUTHOR'S PREFACE.

Early in the present century the method generally adopted for polishing furniture was by rubbing with beeswax and turpentine or with linseed-oil. That process, however, was never considered to be very satisfactory, which fact probably led to experiments being made for the discovery of an improvement. The first intimation of success in this direction appeared in the *Mechanic's Magazine* of November 22, 1823, and ran as follows: "The Parisians have now introduced an entirely new mode of polishing, which is called *plaque*, and is to wood precisely what plating is to metal. The wood by some process is made to resemble marble, and has all the beauty of that article with much of its solidity. It is even asserted by persons who have made trial of the new mode that water may be spilled upon it without staining it." Such was the announcement of an invention which was destined ultimately to become a new industry.

The following pages commence with a description of the art of French Polishing in its earliest infancy, care having been taken by the Author, to the best of his ability, to note all the new processes and manipulations, as well as to concisely and perspicuously arrange and describe the various materials employed, not only for French polishing but for the improving and preparation of furniture woods, a matter of great importance to the polisher. The arts of Staining and Imitating, whereby inferior woods are made to resemble the most costly, are also fully treated, as well as the processes of Enamelling, both in oil-varnishes and French polish, together with the method of decorating the same. The condition of the art of polishing in America is dwelt upon, and various interesting articles written by practical polishers in the States, which appeared in their trade journal, *The Cabinet-maker*, have been revised and printed in this work.

A number of valuable recipes, and other instructive matter, useful alike to the amateur and to the practical workman, are also given.

CONTENTS.

vii

CHAPTER I.

THE IMPROVING AND PREPARATION OF FURNITURE WOODS.

Improving
Matching
Painting
Dyed Polishes

CHAPTER II.

STAINS AND IMITATIONS.

Imitation Mahogany
Imitation Rosewood
Imitation Walnut
Imitation Ebony
Imitation Oak
Imitation Satin-wood
A Blue Stain
A Green Stain
A Purple Stain
A Red Stain
Imitation Purple-wood Stain
Chemicals used in Staining
Process of Staining
Ready-made Wood Stains

CHAPTER III.

FRENCH POLISHING.

The Polish Used
Rubbers
Position
Filling-in
Applying the Polish
Spiriting-off
Prepared Spirits
Antique Style
Dull or Egg-shell Polish
Polishing in the Lathe

CHAPTER IV.

CHEAP WORK.

Glazing
Stencilling
Charcoal Polishing

CHAPTER V.

RE-POLISHING OLD WORK

CHAPTER VI.

SPIRIT VARNISHING.

Varnishes
Brushes and Pencils
Mode of Operation
East Indian Varnishes

CHAPTER VII.

GENERAL INSTRUCTIONS.

Remarks on Polishing
The Polishing Shop

CHAPTER VIII.

ENAMELLING.

Materials
Tools
Mode of Operation
Polishing
Another Process
Decorations

CHAPTER IX.

ix

AMERICAN POLISHING PROCESSES.

Use of Fillers
Making Fillers
Japan of the Best Quality
Fillings for Light Woods
Another for Light Woods
For Mahogany or Cherry Wood
For Oak Wood
For Rosewood
For Black Walnut (1)
"   (2)
An Oil Colour for Black Walnut (3)
Finishing
Black Walnut Finishing
Finishing Veneered Panels, etc.
Light Woods (Dead Finish)
Mahogany or Cherry Wood
Oak
Rosewood, Coromandel, or Kingwood (a Bright Finish)
Walnut
Finishing Cheap Work
With One Coat of Varnish
Wax Finishing
A Varnish Polish
With Copal or Zanzibar Varnish
Polishing Varnish

An American Polish Reviver

CHAPTER X.

MISCELLANEOUS RECIPES.

Oil Polish
Wax Polish
Waterproof French Polish
Varnish for Musical Instruments
French Varnish for Cabinet-work
Mastic Varnish
Cabinet-maker's Varnish
Amber Varnish
Colourless Varnish with Copal
Seedlac Varnish
Patent Varnish for Wood or Canvas
Copal Varnish
Carriage Varnish
Transparent Varnish0 x
Crystal Varnish for Maps, etc.
Black Varnish
Black Polish
Varnish for Iron
Varnish for Tools
To Make Labels Adhere to a Polished Surface
To Remove French Polish or Varnish from Old Work
Colouring for Carcase Work
Cheap but Valuable Stain for the Sap of Black Walnut
Polish (American) for Removing Stains, etc., from Furniture
Walnut Stain to be used on Pine and White-wood

Rosewood Stain
Rosewood Stain for Cane Work, etc.
French Polish Reviver
Morocco Leather Reviver
Hair-cloth Reviver
To Remove Grease Stains from Silks, Damasks, Cloth, etc.
To Remove Ink Stains from White Marble

CHAPTER XI.

MATERIALS USED.

Alkanet-root

Madder-root

Red-sanders

Logwood

Fustic

Turmeric

Indigo

Persian Berries

Nut-galls

Catechu

Thus

Sandarach

Mastic

Benzoin

Copal

Dragon's Blood

Shellac

Amber

Pumice-stone

Linseed-oil
Venice Turpentine
Oil of Turpentine
Methylated Spirits

FRENCH POLISHING
1 AND
ENAMELLING.

CHAPTER I.

THE IMPROVING AND PREPARATION OF FURNITURE WOODS.

For a French polisher to be considered a good workman he should, in addition to his ordinary ability to lay on a good polish, possess considerable knowledge of the various kinds of wood used for furniture, as well as the most approved method of bringing out to the fullest extent their natural tones or tints; he should also be able to improve the inferior kinds of wood, and to stain, bleach, or match any of the fancy materials to which his art is applied, in a manner that will produce the greatest perfection. The following information is given to facilitate a thorough knowledge of the above processes.

2

Improving.—Iron filings added to a decoction of gall-nuts and vinegar will give to ebony which has been discoloured an intense black, after brushing over once or twice. Walnut or poor-coloured rosewood can be improved by boiling half an ounce of walnut-shell extract and the same quantity of catechu in a quart of soft-water, and applying with a sponge. Half a pound of walnut husks and a like quantity of oak bark boiled in half a gallon of water will produce much the same result. Common mahogany can be improved by rubbing it with powdered red-chalk (ruddle) and a woollen rag, or by first wiping the surface with liquid ammonia, and red-oiling afterwards. For a rich mild red colour, rectified spirits of naphtha, dyed with camwood dust, or an oily decoction of alkanet-root. Methylated spirits and a small quantity of dragon's blood will also produce a mild red. Any yellow wood can be improved by an alcoholic solution of Persian berries, fustic, turmeric, or gamboge. An aqueous decoction of barberry-root will serve the same purpose.

Birch when preferred a warm tint may be sponged with oil, very slightly tinted with rose-madder or Venetian red; the greatest care should be used, or it will be rendered unnatural in appearance by becoming too red. Maple which is of a dirty-brown colour, or of a cold grey tint, and mahogany, ash, oak, or any of the light-coloured woods, can be whitened by the bleaching fluid (see "Matching"). Numerous materials [3] may be improved by the aid of raw linseed-oil mixed with a little spirits of turpentine. Artificial graining may be given to various woods by means of a camel-hair pencil and raw oil; two or three coats should be given, and after standing for some time the ground should have one coat of oil much diluted with spirits of turpentine, and then rubbed off.

Matching.—Old mahogany furniture which has been repaired may be easily matched by wiping over the new portions with water in which a nodule of lime has been dissolved, or by common soda and water. The darkeners for general use are dyed oils, logwood, aquafortis, sulphate of iron, and nitrate of silver, with exposure to the sun's rays. For new furniture in oak, ash, maple, etc., the process of matching requires care and skill. When it is desirable to render all the parts in a piece of furniture of one uniform tone or tint, bleach the dark parts with a solution of oxalic acid dissolved in hot water (about two-pennyworth of acid to half a pint of water is a powerful solution); when dry, if this should not be sufficient, apply the white stain (see pp. 11, 12) delicately toned down, or the light parts may be oiled. For preserving the intermediate tones, coat them with white polish by means of a camel-hair pencil. On numerous woods, carbonate of soda and bichromate of potash are very effective as darkeners, as are also other preparations of an [4] acid or alkaline nature, but the two given above are the best.

A good way of preparing these darkeners, says the "French Polisher's Manual," an excellent little work published in Perth some years since, is to procure twopennyworth of carbonate of soda in powder, and dissolve it in half a pint of boiling water; then have ready three bottles, and label them one, two, three. Into one put half the solution, and into the other two half a gill each; to number two add an additional gill of water, and to number three two gills. Then get the same quantity of bichromate of potash, and prepare it in a like manner; you will then have six staining fluids for procuring a series of brown and dark tints suitable for nearly all classes of wood.

The bichromate of potash is useful to darken oak, walnut, beech, or mahogany, but if applied to ash it renders it of a greenish cast. If a sappy piece of walnut should be used either in the solid or veneer, darken it to match the ground colour, and then fill in the dark markings with a feather and the black stain (see pp. 10, 11). The carbonate solutions are generally used for dark surfaces, such as rosewood represents, and a still darker shade can be given to any one by oiling over after the stain is dry. The better way of using these chemical stains is to pour out into a saucer as much as will serve the purpose, and to apply it quickly with a sponge rubbed rapidly and evenly over the surface, and rubbed off dry 5 immediately with old rags. Dark and light portions, between which the contrast is slight, may be made to match by varnishing the former and darkening the latter with oil, which should remain on it sufficiently long; by this means the different portions may frequently be made to match without having recourse to bleaching or staining.

Painting.—The next process is painting. It frequently happens in cabinet work that a faulty place is not discovered until after the work is cleaned off; the skill of the polisher is then required to paint it to match the other. A box containing the following colours in powder will be found of great utility, and when required for use they should be mixed with French polish and applied with a brush. The pigments most suitable are: drop black, raw sienna, raw and burnt umber, Vandyke brown, French Naples yellow (bear in mind that this is a very opaque pigment), cadmium yellow, madder carmine (these are expensive), flake white, and light or Venetian red; before mixing, the colours should be finely pounded. The above method of painting, however, has this objection for the best class of furniture, that the effects of time will darken the body of the piece of furniture, whilst the painted portion will remain very nearly its original colour. In first-class work, therefore, stained polishes or varnishes should be applied instead of these pigments. 6

Dyed Polishes.—The methods of dyeing polish or varnish are as follows: for a red, put a little alkanet-root or camwood dust into a bottle containing polish or varnish; for a bright yellow, a small piece of aloes; for a yellow, ground turmeric or gamboge; for a brown, carbonate of soda and a very small quantity of dragon's blood; and for a black, a few logwood chips, gall-nuts, and copperas, or by the addition of gas-black.

The aniline dyes (black excepted) are very valuable for dyeing polishes, the most useful being Turkey-red, sultan red, purple, and brown. A small portion is put into the polish, which soon dissolves it, and no straining is required. The cheapest way to purchase these dyes is by the ounce or half-ounce. The penny packets sold by chemists are too expensive, although a little goes a long way.

CHAPTER II.

STAINS AND IMITATIONS.

In consequence of the high price demanded for furniture made of the costly woods, the art of the chemist has been called into requisition to produce upon the inferior woods an analogous effect at a trifling expense. The materials employed in the artificial colouring of wood are both mineral and vegetable; the mineral is the most permanent, and when caused by chemical decomposition within the pores it acts as a preservative agent in a greater or less degree. The vegetable colouring matters do not penetrate so easily, probably on account of the affinity of the woody fibre for the colouring matter, whereby the whole of the latter is taken up by the parts of the wood with which it first comes into contact. Different intermediate shades, in great variety, may be obtained by combinations of colouring matters, according to the tint desired, and the ideas of the stainer. The processes technically known as "grounding and ingraining" are partly chemical and partly mechanical, and are designed to teach the various modes of operation whereby the above effects can be produced. We will commence with

Imitation Mahogany.—Half a pound of madder-root, and two ounces of logwood chips boiled in a gallon of water. Brush over while hot; when dry, go over it with a solution of pearlash, a drachm to a pint. Beech or birch, brushed with aquafortis in sweeping regular strokes, and immediately dried in front of a good fire, form very good imitations of old wood. Venetian red mixed with raw linseed-oil also forms a good stain.

The following is a method in common use by French cabinet-makers. The white wood is first brushed over with a diluted solution of nitrous acid; next, with a solution made of methylated spirits one gill, carbonate of soda three-quarters of an ounce, and dragon's blood a quarter of an ounce; and a little red tint is added to the varnish or polish used afterwards. Black American walnut can be made to imitate mahogany by brushing it over with a weak solution of nitric acid.

Imitation Rosewood.—Boil half a pound of logwood chips in three pints of water until the decoction is a very dark red; then add an ounce of salt of tartar. Give the work three coats boiling hot; then with a graining tool or a feather fill in the dark markings with the black stain. 9 A stain of a very bright shade can be made with methylated spirits half a gallon, camwood three-quarters of a pound, red-sanders a quarter of a pound, extract of logwood half a pound, aquafortis one ounce. When dissolved, it is ready for use. This makes a very bright ground. It should be applied in three coats over the whole surface, and when dry it is glass-papered down with fine paper to a smooth surface, and is then ready for graining. The fibril veins are produced by passing a graining tool with a slight vibratory motion, so as to effect the natural-looking streaks, using the black stain. A coat of the bichromate of potash solution referred to on page 4 will make wildly-figured mahogany have the appearance of rosewood.

Imitation Walnut.—A mixture of two parts of brown umber and one part of sulphuric acid, with spirits of wine or methylated spirits added until it is sufficiently fluid, will serve for white wood. Showy elm-wood, after being delicately darkened with the bichromate solution No. 1, page 4, will pass for walnut; it is usually applied on the cheap loo-table pillars, which are made of elm-wood. Equal portions of the bichromate and carbonate solutions (see page 4), used upon American pine, will have a very good effect.

Another method for imitating walnut is as follows: One part (by weight) of walnut-shell extract is dissolved in six parts of soft-water, and 10 slowly heated to boiling until the solution is complete. The surface to be stained is cleaned and dried, and the solution applied once or twice; when half-dry, the whole is gone over again with one part of chromate of potash boiled in five parts of water. It is then dried, rubbed down, and polished in the ordinary way.

The extract of walnut-shells and chromate of potash are procurable at any large druggist's establishment. A dark-brown is the result of the action of copper salts on the yellow prussiate of potash; the sulphate of copper in soft woods gives a pretty reddish-brown colour, in streaks and shades, and becomes very rich after polishing or varnishing. Different solutions penetrate with different degrees of facility. In applying, for instance, acetate of copper and prussiate of potash to larch, the sap-wood is coloured most when the acetate is introduced first; but when the prussiate is first introduced, the heart-wood is the most deeply coloured. Pyrolignite of iron causes a dark-grey colour in beech, from the action and tannin in the wood on the oxide of iron; while in larch it merely darkens the natural colour. Most of the tints, especially those caused by the prussiates of iron and copper, are improved by the exposure to light, and the richest colours are produced when the process is carried out rapidly.

Imitation Ebony.—Take half a gallon of strong vinegar, one pound of extract of logwood, 11 a quarter of a pound of copperas, two ounces of China blue, and one ounce of nut-gall. Put these into an iron pot, and boil them over a slow fire till they are well dissolved. When cool, the mixture is ready for use. Add a gill of iron filings steeped in vinegar. The above makes a perfect jet black, equal to the best black ebony. A very good black is obtained by a solution of sulphate of copper and nitric acid; when dry, the work should have a coat of strong logwood stain.

Imitation Oak.—To imitate old oak, the process known as "fumigating" is the best. This is produced by two ounces of American potash and two ounces of pearlash mixed together in a vessel containing one quart of hot water.

Another method is by dissolving a lump of bichromate of potash in warm water; the tint can be varied by adding more water. This is best done out of doors in a good light. Very often in sending for bichromate of potash a mistake is made, and chromate of potash is procured instead; this is of a yellow colour, and will not answer the purpose. The bichromate of potash is the most powerful, and is of a red colour. A solution of asphaltum in spirits of turpentine is frequently used to darken new oak which is intended for painter's varnish, or a coating of boiled oil.

Another method of imitating new oak upon any of the inferior light-coloured woods is to give the surface a coat of Stephens's satin-wood stain, and 12 to draw a soft graining-comb gently over it, and when the streaky appearance is thus produced a camel-hair pencil should be taken and the veins formed with white stain. This is made by digesting three-quarters of an ounce of flake white (subnitrate of bismuth), and about an ounce of isinglass in two gills of boiling water; it can be made thinner by adding more water, or can be slightly tinted if desired.

Proficients in staining and imitating can make American ash so like oak that experienced judges are frequently deceived, the vein and shade of the spurious wood looking nearly as natural as the genuine. After the veining is done, it should be coated with white hard varnish, made rather thin by adding more spirits, after which the ground can be delicately darkened if required.

Imitation Satin-wood.—Take methylated spirits one quart, ground turmeric three ounces, powdered gamboge one and a-half ounces. This mixture should be steeped to its full strength, and then strained through fine muslin, when it will be ready for use. Apply with a sponge, and give two coats; when dry, glass-paper down with fine old paper. This makes a good imitation for inside work. By the addition of a little dragon's blood an orange tint can be produced. A yellow colour can also be given to wood by boiling hot solutions of turmeric, Persian berries, fustic, etc. but the colour is very fugitive. A more per13 manent colour results from nitric acid, and last of all by the successive introduction of acetate of lead and chromate of potash. Sulphate of iron also stains wood of a yellowish colour when used as a preservative agent, so much so, that the use of corrosive sublimate is recommended for this purpose when it is desirable to preserve the light colour.

A Blue Stain.—This dye can be obtained by dissolving East Indian indigo in arsenious acid, which will give a dark blue. A lighter blue can be obtained by hot solutions of indigo, of sulphate of copper, and by the successive introduction of pyrolignite of iron and prussiate of potash.

A Green Stain.—Dissolve one ounce of Roman vitriol in a quart of boiling water, to which is added one ounce of pearlash; the mixture should then be forcibly agitated, and a small quantity of pulverised yellow arsenic stirred in. A green is also the result of successive formations in the pores of the wood of a blue and a yellow as above indicated, and by a hot solution of acetate of copper in water. A yellowish green may be obtained by the action of copper salts on the red prussiate of potash.

A Purple Stain.—Boil one pound of logwood chips in three quarts of water, until the full 14 strength is obtained; then add four ounces of pearlash and two ounces of powdered indigo. When these ingredients are thoroughly dissolved, it is ready for use, either hot or cold. A purple is also obtained by a boiling hot solution of logwood and Brazil-wood, one pound of the former and one quarter of a pound of the latter to a gallon of water.

A Red Stain.—Methylated spirits one quart, Brazil-wood three ounces, dragon's blood half an ounce, cochineal half an ounce, saffron one ounce. Steep the whole to its full strength, and strain. A red can also be produced by macerating red-sanders in rectified spirits of naphtha. An orange-red colour may be obtained by the successive action of bichloride of mercury and iodide of potash, madder, and ammoniacal solutions of carmine.

Imitation Purple-wood Stain.—Grind a piece of green copperas on coarse glass-paper, and mix with polish coloured with red-sanders. This makes a capital purple stain, and is used by French cabinet-makers.

These dyestuffs may be much improved by the addition of a mordant applied after they are dry; this will greatly assist in modifying and fixing the tints and shades which the dyes impart. The best thing for the purpose, in the writer's opinion, is clear ox-gall, which, besides being useful as a mordant, will destroy all unctuous matter. 15

Chemicals used in Staining.—It may perhaps be useful here to give the common or popular names of the chemicals employed in the operations of staining and imitating, as few polishers know them by the scientific names used by chemists:—

Nitric acid is but another phrase for aquafortis.
Sulphuric acid, for oil of vitriol.
Ammonia, for spirits of hartshorn.
Sulphate of magnesia, for Epsom salts.
Nitrate of potass, for sal prunelle.
Chlorine, for aqua regia.
Sulphate of copper, for blue vitriol.
Subborate of soda, for borax.
Superoxalate of potass, for salts of sorrel.
Hydrochlorate of ammonia, for sal ammoniac.
Subnitrate of bismuth, for flake white.
Acetic acid, for vinegar.
Acetate of lead, for sugar of lead.
Sulphate of lime, for gypsum.
Carbonate of potass, for pearlash.
Bitartrate of potass, for cream of tartar.
Nitrate of silver, for lunar caustic.
Supercarbonate of iron, for plumbago.
Cyanide of iron, for Prussian blue.
Subacetate of copper, for common verdigris.
Susquecarbonate of ammonia, for sal volatile.
Alcohol, for pure spirit.
Sulphate of iron, for green copperas.
Sulphate of zinc, for white copperas.

16 **Process of Staining.**—The natural qualities of woods are very variable; so also are the textures of the different sorts usually used for staining. It will be readily perceived that there is no fixed principle upon which certain peculiar tints or shades can be produced with any degree of certainty. In order to arrive at the best results, the stainer is recommended to observe the following rules:—

All dry stuffs are best reduced to powder, when it is possible, before macerating or dissolving them.

All liquids should be strained or filtered before use.

The requisite ingredients should always be tested before a free use is made of them, as the effect produced by a coat of stain cannot be accurately ascertained until it is thoroughly dry.

Amateurs in staining had far better coat twice or thrice with a weak stain than apply a strong one; for if too dark a tint is first obtained it is often irremediable. Flat surfaces will take stain more evenly if a small portion of linseed-oil is first wiped over, well rubbed off, and allowed to dry, then lightly papered down with fine glass-paper. End-way wood which is of a spongy nature should first have a coat of thin varnish, and when dry well glass-papered off. For applying stain a flat hog-hair tool is the best; and for a softener-down a badger-hair tool is used. For mahogany shades and tints a mottler will be found of service, as will also a soft piece of Turkey sponge. For oak, 17 the usual steel graining-comb is employed for the streaking, and for veining badger sash-tools and sable pencils.

Ready-made Wood Stains.—There are numerous stains suitable for common work in the market obtainable at a small cost by residents in London, but it is cheaper for those who reside in country towns to make their own, if only a small quantity is required. The principal makers of wood stains are H. C. Stephens, of 191, Aldersgate-street, E.C., and Jackson, 213, Union-street, Southwark, S.E. These makers prepare stains in a liquid state, and also in powders for oak, walnut, mahogany, satin-wood, ebony, and rosewood. The powders are sold in packages at 8s. per lb. or 1s. for two ounces, and are soluble in boiling water. Judson, of 77, Southwark-street, S.E., makes a mahogany powder in sixpenny packets, and any reliable oilman will sell a good black stain at 8d. per quart, or a superior black stain at 1s. 2d. per quart. Fox, of 109, Bethnal Green-road, also prepares stains in a liquid state.18

CHAPTER III.

FRENCH POLISHING.

The Polish Used.—The oil or wax polish was used for all kinds of furniture before the introduction of French polish, the invention of which, as its name implies, is due to French cabinet-makers. It was first introduced into England about seventy years since; some time elapsed, however, before it was brought to a high state of perfection. At first apprentices or porters were entrusted with the polishing, they having been usually called upon to do the wax polishing; but in course of time it was found that its successful adoption implied the possession of considerable skill, and it came to be regarded as an art of no little importance—so much so, that the early polishers who had perfected themselves used to work in a shop with closed doors, lest the secret of their success should be discovered. From that time polishing became a separate branch of the cabinet business.

The following original recipe as first invented has been extracted from a French work, the 19 *Dictionnaire Technologique*, not, however, for its usefulness (it having gone into disuse many years ago), but as a matter of curiosity:—

"*French Polish.*"

Gum sandarach	14 ounces 2 drachms
Gum mastic in drops	7 " 2 "
Shellac (the yellower the better)	14 " 2 "
Alcohol of 0.8295 specific gravity	3 quarts and 1 pint.

"Pound the resinous gums, and effect their solution by continued agitation, without the aid of heat; if the woods are porous, add seven ounces one drachm of Venice turpentine. If an equal weight of ground glass be added, the solution is more quickly made, and is also otherwise benefited by it. Before using, the wood should be made to imbibe a little linseed-oil, the excess of which should be removed by an old flannel."

Notwithstanding the improvement made upon the old processes by this new method, it was by no means considered to be perfect, for the polish was found to impart its brown tinge to the light-coloured woods, especially in marqueterie work, and to deteriorate their appearance. It will be readily seen that it was a great desideratum among polishers to render shellac colourless, as, with the exception of its dark-brown hue, it possesses all the properties essential to a good polish or spirit varnish in a higher degree than any of the other resins.

In 1827 the Society of Arts came forward with its valuable aid and offered a premium of a gold 20 medal, or thirty guineas, "for a polish or varnish made from shell or seed-lac, equally hard, and as fit for use in the arts as that at present prepared from the above substance, but deprived of its colouring matter." After numerous experiments, this long-felt want was perfectly attained by Dr. Hare, who was awarded the premium. His method was as follows: "Dissolve in an iron kettle one part of pearlash in about eight parts of water; add one part of shell or seed-lac, and heat the whole to ebullition. When the lac is dissolved, cool the solution, and impregnate it with chlorine till the lac is all precipitated. The precipitate is white, but its colour deepens by washing and consolidation; dissolved in alcohol, lac, bleached by the process above mentioned, yields a polish or varnish which is as free from colour as any copal varnish." At the present time shellac is bleached by filtration over animal charcoal.

Numerous experiments were afterwards made in the manufacture of polishes; several chemists devoted their attention to its manufacture, and an improved polish was soon produced which was used for a number of years. The following are its proportions:—

Shellac	14 ounces.
Sandarach	¾ ounce.
White resin	¾ "
Benzoin	¾ "
Gum thus	¾ "
O.P. finishing spirit	½ a gallon.

The[21] "filling-in" processes also began to be used, which effected a considerable saving in the quantity of polish usually required, and in consequence of the expensiveness of spirits of wine rectified spirits of naphtha was used as a substitute for the making of polishes, etc.; but it was discovered that its continued use soon affected the eyesight of the workmen, and it had to be abandoned, the methylic alcohol, pyroxylic spirit, or wood spirit, as it has been differently called, taking its place. This was first discovered by Mr. Philip Taylor in 1812, and is obtained by distilling wood. Messrs. Dumas & Peligot, after analysing it, determined that it contained 37·5 per cent. of carbon, 12·5 per cent. of hydrogen, and 50 per cent. of oxygen. When pure, it remains clear in the atmosphere; but for the sake of economy it is often employed in the manufacture of other compounds called methylated. This spirit began to be much used in the manufacture of polishes and varnishes in the year 1848, and has continued to be much used ever since.

The wonderful improvements which have been effected in polishes since their first introduction obviously prove that they have now arrived at a very high point of perfection, and polishing is now justly acknowledged, both by skilled artisans and connoisseurs, to be an important decorative art. French polish or varnish at the present time can easily be obtained at most chemists or oil shops, or direct from the manufacturers, 22 amongst whom may be mentioned Mr. W. Urquhart, 327, Edgware-road, W.; Messrs. Turner & Sons, 7 to 9, Broad-street, Bloomsbury, W.C.; Messrs. William Fox & Son, Bethnal Green-road, E.; Mr. G. Purdom, 49, Commercial-road, E.

The London prices are: Best French polish, 5s. 6d. per gallon; best white polish, 9s. per gallon; brown or white hard varnish, 8s. per gallon; patent glaze, 10s. per gallon; methylated spirits, 3s. 3d. per gallon. For those who prefer to make their own, the following will be found an excellent recipe:—

12 ounces of orange shellac.
1 ounce of benzoin.
1 ounce of sandarach.
½ gallon of methylated spirits.

Pound the gums well before mixing with the spirit, as this will hasten their dissolution. White polish for white wood and marqueterie work should be made with bleached shellac instead of the above. In making polishes or varnishes, the mixture will frequently require shaking until dissolved.

Rubbers.—In commencing to polish, the materials required are old flannel for the rubbers and clean old linen or cotton rags for the coverings, the softer the better; some polishers, however, prefer white wadding for rubbers instead of flannel. Rubbers for large surfaces 23 are usually made of soft old flannel, firmly and compactly put together somewhat in the form of a ball, and the more they possess softness and compactness, and are large and solid, the more quickly and satisfactorily will they polish extensive surfaces. Small pliable rubbers are usually employed for chairs or light frame-work. Perhaps for a beginner a rubber made of old flannel may be best, as it takes some little practice to obtain the necessary lightness of hand.

The rubber for "spiriting-off" should be made up from a piece of old flannel or woollen cloth, and covered with a piece of close rag, doubled. Carefully fold the rag and screw it round at the back to make it as firm as possible, and sprinkle some spirit on the face of it just as it is covered; then give it two or three good smacks with the palm of the hand, and begin by laying on as lightly as possible for the first few strokes and gradually increase the pressure as the rubber gets dry; then take off the first fold of the cover and work it perfectly dry.

The rubber should present to the wood you are about to work on a smooth-rounded or convex surface. Have beside you linseed-oil in one receptacle, and some French polish in another. Apply one drop of polish and one drop of oil, and polish with a circular movement—traversing steadily the *whole* surface to be polished, and from time to time, as may be necessary (when the rubber gets sticky and harsh, indicating that the 24 inside needs replenishing with more polish), open and apply more of it, and again draw over the linen cover, holding it tightly to form the convex face proper to do the work. After replenishing, the rubber will probably need a little more oil to help it to work smoothly. Having thus worked on one coat of polish evenly over all the wood until it has got what may best be described as a *full look*, set it aside for two or three hours to sink in and become hard, and when completely dry, lightly paper off with glass-paper (very finest), afterwards dusting the surface to remove any trace of powder, and lay on a second coat of polish in the same method. Then allow twenty-four hours to dry. Another light papering may possibly be needed—dust off as before recommended and let the wood have a *third* coat of polish.

For this third coat a fresh rubber should be made, the inside being sparingly supplied with spirits of wine instead of polish. Put a double fold of linen over it, touch it with one drop of oil, and go very lightly and speedily over the whole work.

In spiriting-off—the object of which is to remove any trace of smudge that may blur the surface unless removed thus by spirit—you should go gently to work, using a very light hand, or you may take the polish off as well, amateurs more especially.

Position.—All work should be placed in an 25 easy and accessible position while it is being polished, so that the eye may readily perceive the effect of the rubber; this will greatly help to relieve the difficulties attending the polishing of a fine piece of furniture. It should also be kept quite firm, so that it cannot possibly move about. The most suitable benches for polishers are the ordinary cabinet-makers' benches, with the tops covered with thick, soft cloths; these cloths should not be fastened down, it being an advantage to be able to remove them when required. When a piece of work too large to be placed upon the bench is in hand, pads will be found useful to rest it upon. These can be made by covering pieces of wood about two feet in length by three inches in width with cloth several times doubled, the work being placed so that a good light falls upon it. All thin panelling should be tacked down upon a board by the edges while polishing.

Filling-in.—The first process the wood usually undergoes is "filling-in." This consists in rubbing into the pores of the wood Russian tallow and plaster of Paris, which have been previously heated and mixed together so as to form a thick paste. For rosewood, or to darken mahogany, a little rose-pink should be added. After well rubbing in, the surface should be cleared from all the surplus paste with the end of the scraper, and then rubbed off with shavings 26 or old rags, and made quite clean. For birch or oak, some use whiting or soft putty moistened with linseed-oil for the filling; this preparation prevents in a great measure the rising of the grain. For white delicate woods, such as sycamore, maple, or satin-wood, plaster of Paris, mixed with methylated spirit, is used. When polishing pine, a coat of Young's patent size (2d. per lb.) is used instead of the above mixtures, and when dry is rubbed down with fine glass-paper.

Some workmen, who regard their modes of filling-in as important secrets, do their work surprisingly quick by the methods here given. The various processes are soon acquired by a little practice, and contribute greatly to the speedy advancement of a smooth and imporous ground, which is the most important point to observe in polishing.

Applying the Polish. — In commencing to use the polish some are provided with a small earthenware dish, into which the polish is poured for wetting the rubbers; while others make a slit in the cork of the polish bottle, and so let it drip on to the rubber; whichever method is adopted, the rubber should not be saturated, but receive just enough to make a smear. Every time after wetting the rubber and putting on the cover it should be pressed upon the palm of the hand, or if a small rubber it can be tested between 27 the thumb and finger. This is an important operation, for by it the polisher can easily tell the exact state of moisture, and at the same time, by the pressure being applied, the moisture is equalised. The tip of the finger should then be just dipped into the linseed-oil, and applied to the face of it; if the rubber should be rather sappy, the greatest care must be used or a coarse streaky roughness will be produced; extreme lightness of hand is required until the rubber is nearly dry. (It would be a serious error to bear heavily on the rubber while the surface is moist; to do so, and to use too much oil on the rubber, are the causes of many failures in polishing.)

In commencing to work, pass the rubber a few times gently and lightly over the surface in the direction of the grain; then rub across the grain in a series of circular movements, all one way, in full and free sweeping strokes, until the rubber is dry. Continue this operation until the pores are filled in, and the surface assumes a satisfactory appearance. It should then be left for about twelve hours; the polish will be well into the wood by that time. The polish should then be carefully rubbed down with No. glass-paper; this will remove the atomic roughness usually caused by the rising of the grain during the sinking period. In flat-surface work a paper cork can be used, and the rubbing lightly and regularly done in a careful manner, so as to avoid rubbing through the outer skin, especially at the 28 edges and corners, or the work will be irremediably defaced.

The woods which possess a rising grain are well known to polishers; these are the light-coloured woods with a coarse grain, viz., ash, birch, and oak. This rising of the grain can in a great measure be prevented by passing a damp sponge over the work before commencing the polishing, allowing it to dry, and papering it. After the rubbing or smoothing-down process is finished, the work should be well dusted; the polishing can then recommence. The above operation must be again repeated with a rotary motion and gradually increased pressure as the rubber gets dry, and finished by lighter rubbings the way of the grain; this will remove any slight marks that may be occasioned by the circular movements of the rubber.

Working too long on any one part is to be avoided, nor should the rubber be allowed to stick even for an instant, or it will pull the coating of polish off to the bare wood. The rubber should be covered with a clean part of the rag as soon as a shiny appearance becomes apparent upon it, or at each time of damping, and less oil should be used towards the end of the operation, so as to gradually clear it all off from the surface. Rubber marks can be removed by rubbing in a direction the reverse of the marks with a half-dry rubber and increased pressure. When the work has received a sufficient body, in finishing the drying of 29 the last rubber, ply it briskly the way of the grain to produce a clean dry surface for the spiriting-off.

The following is the method usually employed on fine carved or turned work when finished in the best style. In the first place it is embodied with polish, using a small rubber for the operation, after which it should have one coat of shellac (two parts by weight of shellac to one of spirits) applied with a brush, and when dry it should be carefully smoothed down with flour paper, the utmost caution being observed in dealing with the sharp edges, or the carving will be spoiled. Then it is embodied with polish again, and one coat of glaze applied with the greatest care. A few hours should be allowed to harden, and then finished off with a rubber slightly damped with thin polish. This is an expensive method, but it will pay in appearance for all the time bestowed.

For the best class of cabinet and pianoforte work in amboyna or burr-walnut it is advisable not to use linseed-oil on the sole of the rubber when polishing, but the best hog's lard; the reason for this is that these veneers being so extremely thin and porous the oil will quickly penetrate through to the groundwork, softening the glue, and causing the veneers to rise in a number of small blisters. Of course, this is not always the case, but the use of lard instead of oil will be found a good preventative. Lard is also used on the above class of work when it is desirable to preserve the colour of the wood in its natural state. 30

The following method is employed for the best work: Immediately on receiving the job from the cabinet-maker, a good coating of thin, clean glue should be applied with a sponge or brush; this is allowed to dry, and thoroughly harden; it is then cleaned off, using the scraper and glass-paper, cutting it down to the wood. The bodying-in with white polish is the next process, the usual sinking period being allowed; it is again cleaned off, but the scraper this time should not quite reach the wood. Then embody again, and treat in a like manner. In getting up the permanent body, commence with a slight embodying; let this stand, and when the sinking period is over rub it down with a felt rubber and powdered pumice-stone; continue this several times, till the surface presents a satisfactory appearance, and the job is ready for the spiriting process. By this means the wood will retain its natural colour, and a beautiful transparent polish will result, and remain for a number of years. This also is an expensive process, but the result cannot be obtained in any other way.

Spiriting-off.—Most polishers affirm that if an interval of at least a couple of hours elapse between the final embodying and the spiriting-off the brilliancy of the polish will be improved, and remain harder and more durable. The spirit is applied in exactly the same way as the polish, and the same rubber can be used, but it should be 31 covered with more than one fold of the soft linen rag; care should be taken not to make it very wet, or the gum on the surface of the work will be redissolved, and a dulness instead of a brilliancy will result. If the spirit should be very strong, the rubber should be breathed upon before using, and a little more oil taken up; some, however, prefer to mix a little polish with the spirit, while others prefer the spirit to be weakened by exposure to the air for a few hours; experience alone must be the teacher in this particular; but if the spirit should not "bite," as it is termed, all will be well. The last rubber should be worked a little longer than usual, and a trifle quicker, so as to remove the slight greasy moisture on the surface.

The finishing touch is given to the work by a soft rag loosely rolled up and just a few drops of spirit dropped upon it, applied quickly the way of the grain. This will remove every defect, and leave it clear and brilliant. If, in a short time after finishing, the polish becomes dull or rough, it will be owing to too much oil being absorbed in the process and working through the surface, combined with dust. It should be cleaned off first with a soft cloth, damped with a little warm water, and the whole repaired, as at first, with equal parts of polish and spirits mixed together, using the least possible damp of oil to make it finish clear; there is no danger of its happening again. In all cases the work must be rubbed till quite dry, and when nearly so the pressure may be increased.

The rubber for spiriting-off should be made up from a piece of old flannel, and be covered with a piece of old rag. This is preferable to very thin rag, and will give a better finish.

Prepared Spirits.—This preparation is useful for finishing, as it adds to the lustre and durability, as well as removes every defect of other polishes, and it gives the surface a most brilliant appearance.

It is made of half a pint of the very best rectified spirits of wine, two drachms of shellac, and two drachms of gum benzoin. Put these ingredients in a bottle and keep in a warm place till the gum is all dissolved, shaking it frequently; when cold add two teaspoonfuls of the best clear white poppy oil; shake them well together, and it is fit for use.

Antique Style.—For mediæval or old English furniture a dull polish is generally preferred to a French polish, because it has a gloss rather than a brilliant polish, which materially assists in showing up mouldings or carvings to the best advantage; it is also more in character with the work of the Middle Ages. Another advantage is the facility of obtaining a new polish (after being once done) should the first one get tarnished, as the finishing process can be performed without difficulty by any one, and a new polish obtained each time.

On receiving a job which is required to be done in this style, it should be "filled-in" in the usual manner, and afterwards bodied with white polish to a good extent; it is then left for a sinking period (say twelve hours). The work is then carefully rubbed down with powdered pumice-stone and a felt-covered block or rubber, and after well dusting it is ready for finishing. The preparation used for this process is mainly composed of bees'-wax and turpentine (see Wax Polish, page 87), well rubbed in with a piece of felt or a woollen rag, and finished off by rubbing briskly with a very soft cloth or an old handkerchief to produce a gloss.

Dull or Egg-shell Polish.—This is another style of finishing for mediæval work; the process is very simple. In commencing a job to be finished in this style, the process of "filling-in" and "embodying" are first gone through, then a sinking period is allowed, after which it is embodied again, till the work is ready for finishing. All the parts should be carefully examined to see if there is a good coating of polish upon them. This is important, for if the work should be only thinly coated it is liable to be spoiled by rubbing through in the last process. After allowing a few hours for the surface to 34 harden, a pounce bag of powdered pumice-stone should be applied to the work, and a felt-covered rubber used, rubbing down in the direction of the grain until the work is of the desired dulness.

For the cheaper kind of work done in this style, the first process, of course, is the filling-in; then a rubber of wadding is taken and used without a cover, made rather sappy with polish and a few drops of oil added; and after bodying-in with this sufficiently, the work should be stood aside for twelve hours, then rubbed down with some fine worn glass-paper. The embodying is then again commenced, a proper rubber and cover being used; and when sufficient is put on, and while the surface is still soft, the pounce above mentioned should be applied, and rubbed down with a piece of wadding slightly moistened with linseed-oil until the desired dulness appears. This is becoming the fashionable finish for black walnut work.

Polishing in the Lathe.—The lathe is of more use to a polisher than a great many persons outside the trade would imagine. By its aid turned work can be finished in a most superior style, and in less time than by hand. The articles usually done by the lathe are wood musical instruments, such as clarionets, flutes, etc.; also cornice-poles, ends, and mahogany rings, the latter being first placed in a hollow 35 chuck and the insides done, after which they are finished upon the outside on a conical chuck. For table-legs, chair-legs, and all the turnery used in the cabinet-work, it will be found of great advantage to finish the turned parts before the work is put together.

Most of the best houses in the trade finish their work in this way, where all the work is polished out entirely with the rubber. In the first place, the filling-in is done. The band is thrown off the pulley and the work rubbed in; at the same time the pulley is turned round by the left hand. When this is done, the band is replaced and the work cleaned off with rags or shavings, the lathe to be driven with speed to get a clean surface. When applying the polish the lathe should revolve with a very slow motion.

The rubbers best adapted for turned work are made of white wadding, as the hollows and other intricacies can be completely finished out with a soft rubber. The work should first receive a coating of thick shellac, two parts by weight of shellac to one of methylated spirits, and applied with a brush or a soft sponge; after a couple of hours this is nicely smoothed with fine paper, and the "bodying-in" completed with the soft rubber and thin polish. There are numerous hard woods which do not require filling-in, amongst which may be mentioned boxwood, cocus, ebony, etc.; these may be rapidly polished in the lathe, on account of their texture, with the white polish. 36 In spiriting-off a very soft piece of chamois leather (if it is hard and creased it will scratch) should be damped with methylated spirits, then wrung so that the spirit may be equally diffused; the lathe should then be driven at a rapid speed, and the leather held softly to the work. In a few minutes, if a dark wood, a brilliant surface will be produced.

37

CHAPTER IV.

CHEAP WORK.

Glazing.—Glaze is known to the trade under several names, such as slake, finish, and telegraph; it is used only for cheap work, when economy of time is a consideration, and is made as follows: mastic, 1 oz.; benzoin, 5 ozs.; methylated spirit, 5 gills. A superior article can be obtained from G. Purdom, 49, Commercial Road, Whitechapel, E., who is the manufacturer of a "patent glaze."

First give the work a rubber or two of polish after the "filling-in"; it is important to dry the last rubber thoroughly, so that no unctuousness remains upon the surface before applying the glaze, otherwise it will be of no effect. The way to apply it is as follows: Prepare a rubber as for polishing and make it moderately wet, and take only one steady wipe the way of the grain, never going over the same surface twice while wet; and when dry, if one coat is found not to be enough, apply a second in the same manner. For mouldings or the backs and sides of chair-work, this is generally considered to be sufficient. Some 38 polishers will persist in using glaze to a large extent, even on the best-paid work; but it is not recommended, as the surface will not retain its brilliancy for a lengthened period, particularly in hot weather. Nothing is so good for the best class of work as polishing entirely with French polish.

The way of treating small flat surfaces such as the frames of tables, looking-glasses, builders' work, etc., is to first fill in, and give one or two rubbers of polish, drying the last rubber thoroughly; then glaze, and after a period of two or three hours finish with a rubber slightly wetted with thin polish. It is a bad plan to put glaze on newly-spirited work, or to re-apply it on old bodies.

The following is another method for cheap work: A coating of clear size is first given in a warm state (this can be obtained at most oil-shops), and when dry is rubbed down with fine glass-paper, after which a coating of varnish is applied with a sponge or a broad camel-hair brush, giving long sweeping strokes. The tool should be plied with some degree of speed, as spirit varnishes have not the slow setting properties which distinguish those of oil, and care should be taken not to go over the same part twice. When this is thoroughly hard it is nicely smoothed with fine paper, a few rubberfuls of polish is given, and it is then ready for spiriting-off.

Another plan is frequently adopted for cheap 39 work: Make a thin paste with plaster of Paris, suitably tinted and watered, and well rub in across the grain with a piece of felt or old coarse canvas till the pores are all full; any superfluity should be instantly wiped off from the surface before it has time to set. The succeeding processes are papering and oiling. In applying the polish, which should be done immediately after oiling, the rubber should be made rather sappy with thin polish, and worked without oil. During the embodying a pounce-bag containing plaster of Paris is sparingly used; this application tends to fill the pores and also to harden the body of polish on the exterior, but too much should not be used, or it will impart a semi-opaque appearance to the work. This first body is allowed sufficient time to harden; it is then rubbed down lightly with flour paper or old worn No. 1, and then embodied with thicker polish or a mixture of polish and varnish, and the smallest quantity of oil applied to the rubber. When a sufficient body of polish is given to the work, the surface is rubbed very carefully with a lump of moist putty plied in the longitudinal direction of the grain; this will bring up a gloss, and very little spiriting will be required.

Stencilling.—An imitation of marqueterie on light-coloured woods can be obtained by the following method: Cut a stencil pattern in stout cartridge paper (this is best done upon a piece of 40 glass with the point of a sharp penknife), and place it on the centre of a panel or wherever required, and have ready some gas-black mixed with thin polish; apply this with a camel-hair pencil over the cut-out pattern, and when it is removed finish the lines and touch up with a finer tool. The work should be first bodied-in, and when the pattern is dry rubbed down with a piece of hair-cloth (the smooth side down) on a cork rubber to a smooth surface, after which the polishing can be proceeded with until finished. Upon oak this will have the appearance of inlaid work.

Charcoal Polishing.—A method known as "charcoal polishing" is now much used for producing the beautiful dead-black colour which seems to have the density of ebony. Its invention is due to French cabinet-makers. The woods used by them are particularly well adapted for staining black or any other colour, limetree, beech, cherry, pear, soft mahogany, or any wood of a close and compact grain being the woods usually selected.

The first process is to give the work a coating of camphor dissolved in water and made rather strong; this will soon soak into the wood, and immediately afterwards another coat composed of sulphate of iron-water with a few nut-galls added. These solutions in blending penetrate the wood and give it an indelible tinge, and also prevent insects from attacking it. After these coats are 41 dry, rub the surface with a hard brush (an old scrubbing-brush will do) the way of the grain, after which rub the flat parts with natural stick charcoal, and the carved or indented portions with powdered charcoal; the softest portion of the charcoal only should be used, because if a single hard grain should be applied it would seriously damage the surface. The workman should have ready at the same time a preparation of linseed-oil and essence of turpentine (linseed-oil one gill, and essence of turpentine one teaspoonful), a portion of which should be freely taken up with a piece of soft flannel and well rubbed into the work. These rubbings with the preparation and charcoal several times will give the article of furniture a beautiful dead-black colour and polish. This method of polishing is applied to the black-and-gold furniture, cabinets, etc., in imitation of ebony.

Another good black polish is obtained by gas-black being applied to the rubber after wetting with French polish, the cover being then put on and worked in the usual manner.

These black polishes should not be applied if there are coloured woods in the piece of furniture. Should the work be already dyed black, or in black veneers, it is best to use white polish, which will greatly help to preserve the transparent density of the dye.

CHAPTER V.

RE-POLISHING OLD WORK.

If the piece of furniture requiring to be re-polished should be in bad condition, it is best to clean off thoroughly, using the liquid ammonia (see page 94), or by the scraper and glass-paper. The indentations may be erased by dipping into hot water a piece of thick brown paper three or four times doubled and applying it to the part; the point of a red-hot poker should be immediately placed upon the wet paper, which will cause the water to boil into the wood and swell up the bruise; the thickness of the paper prevents the wood from being scorched by the hot poker. After the moisture is evaporated, the paper should be again wetted if required. If only shallow dents, scratches, and broken parts of the polish present themselves, carefully coat them two or three times with a thick solution of shellac, and when the last coating becomes hard carefully paper down with a piece of old glass-paper and a cork rubber.

If the surface should be in good condition, it 43 is necessary only to remove the viscid rust; this is done by friction with a felt-covered rubber and pure spirits of turpentine; by this means the polish remains unsullied. If the surface should not be in very good condition, a flannel should be used smeared with a paste of bathbrick-dust and water, or a paste made of the finest emery flour and spirits of turpentine. After cleansing, and before the polish is applied, it is a good plan to just moisten the surface with raw linseed-oil; this will cause the old body to unite with the new one.

In order to carry out the process of re-polishing with facility, it is necessary to disunite all the various parts, such as panels, carvings, etc., before commencing the operation. The polish is applied in the usual manner, and when a good body is laid on the work should be set aside for twelve hours, after which it can be finished. It should be particularly observed that in polishing no job should be finished immediately after the rubbing-down process; a sinking period should always be allowed. If the work should be immediately finished, the consequences are that in a few hours all the marks and scratches of the paper, etc., will be discernible, and the polished surface will present a very imperfect appearance, although looking perfect when first finished.

Holes and crevices may be well filled up with a cement made in the following manner: In a large iron spoon place a lump of beeswax about 44 the size of a walnut, a pinch of the pigments mentioned on page 5, according to the colour required, a piece of common rosin the size of a nut, and a piece of tallow as large as a pea;

melt, and it is ready for use. Some add a little shellac, but much will make it very brittle. A similar substance to the above can be bought at the French warehouses.

45

CHAPTER VI.

SPIRIT VARNISHING.

Most polishers are agreed that to obtain a good surface with varnish it is necessary to give the work, where it is possible to do so, a rubberful of polish first, and to thoroughly dry the rubber; but in most carved work the surface is not accessible, and the brush must be used. Sometimes the carving is extremely coarse, and with an open porous grain, in which case it is best to oil it first and then to fine-paper it down; by this process a thin paste is formed by the attrition, which materially assists in filling up the pores. Before commencing to use the varnish have ready an earthenware dish or box,—one of the tins used for the preserved meats or fish will answer the purpose,—with two holes drilled so that a piece of wire can be fastened diametrically across the top; this is called a "regulator," and when the brush is passed once or twice over this it prevents an unnecessary quantity of varnish being transferred to the work. 46

Varnishes.—The ingredients for making varnish are very similar to those for making polish, but the proportions are somewhat different. Furniture varnish consists of two kinds, viz.: the brown-hard and the white-hard; the former is used for dark woods, such as mahogany, walnut, rosewood, etc.; whilst the latter is used for the light-coloured woods, in conjunction with the white polish. A few years since the brown-hard varnish was made from these ingredients:

1 gallon of methylated spirit,
40 ozs. of shellac,
4 ozs. of rosin,
5 ozs. of benzoin,
2 ozs. of sandarach,

2 ozs. of white rosin.

The brown-hard varnish which is used at the present time is made differently, and produces a better result; it is made from the following:

1 gallon of methylated spirit,
32 ozs. of shellac,
8 ozs. of rosin,
8 ozs. of benzoin.

The white-hard or transparent varnish for white wood is made with

1 gallon of methylated spirit,
32 ozs. of bleached shellac,
24 ozs. of gum sandarach.

In making either polishes or varnishes, all the 47 gums should be first pounded and reduced to powder before mixing with the spirit, and when mixed they should be occasionally well shaken or stirred, so as to hasten their dissolution.

Brushes and Pencils. — The brushes used for varnishing are either flat, in tin, or round, tied firmly to the handle, and made of camel's-hair; but the small white bristle-tools and red-sable pencils will frequently be found of service in coating delicate carving, or turned work. Varnish brushes can be obtained from a quarter of an inch to four inches and upwards in width; the most useful brush, however, for general use is about an inch wide. It is important that brushes should be cleaned in spirits immediately after use, for if laid by in varnish they lose their elasticity and are soon spoiled; but if this preservative principle is ever neglected, the hardened brush should be soaked in methylated spirit, and if wanted for immediate use the spirit will soften the varnish quicker if made luke-warm. The spirit should be gently pressed out by the finger and thumb. All varnish brushes when not in use should be hung up, or kept in such a position that they do not rest upon their hairy ends, either in a box or tin free from dust.

Mode of Operation.—It is usual in varnishing to give the work three coats, and always allow each coat to dry thoroughly before applying the next. It should be noted that spirit varnishes begin to dry immediately they are laid on; therefore, on no account should they be touched with the brush again whilst wet, or when dry they will present a rough surface. Always ply the brush quickly, and never go over a second time. When giving the first or second coats it is unimportant how they are applied, whether across the grain or with the grain, but the finishing coat should always be with the grain. If the varnish should appear frothy when laid on, it is of no consequence, as it will dry smooth if equally and evenly applied before a good fire or in a warm atmosphere.

Coloured varnishes can be made in exactly the same manner as coloured polishes (see page 6). The beautiful glossy black varnishes so admired on Indian cabinet-work, specimens of which can be seen at the Indian Museum, are very difficult to obtain in England, but a description of them may be interesting.

East Indian Varnishes.—The Sylhet varnish is composed of two parts of the juice of the bhela (the tree which bears the marking nuts of India), and one part of the juice of the jowar. The articles varnished with it at Sylhet are of the most beautiful glossy black; and it seems equally fitted for varnishing iron, leather, paper, wood, or stone. It has a sort of whitish-grey colour when first taken out of the bottle, but in a few minutes it becomes perfectly black by exposure to the air. In the temperature of this country it is too thick to be laid on alone; but it may be rendered more fluid by heat. In this case, however, it is clammy, and seems to dry very slowly. When diluted with spirits of turpentine, it dries more quickly; but still with less rapidity than is desirable.

The *tsitsi*, or Rangoon varnish, is less known than the Sylhet varnish. It is probably made from the juice of the bhela alone. It appears to have the same general properties as the Sylhet varnish, but dries more rapidly. The varnish from the *kheeso*, or varnish-tree, may be the same as the Rangoon varnish, but is at present considered to be very different. The kheeso grows particularly in Kubboo, a valley on the banks of the Ningtee, between Munnipore and the Burman empire. It attains to such a large size, that it affords planks upwards of three feet in breadth, and in appearance and grain is very like mahogany. A similar tree is found in great abundance and perfection at Martaban.

A poisonous vapour exhales from several of the Indian varnishes, especially from that of Sylhet, and is apt to produce over the whole skin inflammations, swellings, itchings, and pustules, as if the body had been stung by a number of wasps. Its 50 effects, however, go off in a few hours. As a preventative the persons who collect the varnish, before going to work, smear their faces and hands with greasy matter to prevent the varnish poison coming into contact with their skin.

51

CHAPTER VII.

GENERAL INSTRUCTIONS.

Remarks on Polishing.—Amateurs at French polishing will be more successful on a large surface than a small one.

When polishing, the rubber-cloth should be changed occasionally, or the brightness will not remain when finished.

A most efficacious improver of many kinds of woods is raw linseed-oil mixed with a little rectified spirits of turpentine.

French polish can be tinted a light-red with alkanet-root, and a dark-red with dragon's blood.

A good Turkey sponge is capable of spreading either stain or varnish more smoothly than a camel's-hair brush on a flat surface.

The sub-nitrate of bismuth mentioned on p. 12 is beginning to supersede oxalic acid for bleaching processes.

Thin panels for doors should be securely tacked down to a level board, and polished with a 52 large round flannel rubber having a very flat sole. Fret-work panels should have all the edges entirely finished with varnish before they undergo the above operation. To get a good polish upon a full-fret panel is considered by polishers to be the most difficult part in the work, on account of the extreme delicacy and frangibility of the work and the great carefulness required.

Soft spongy wood may be satiated by rubbing a sponge well filled with polish across the grain until it becomes dry.

In polishing a very large surface, such as a Loo-table top or a wardrobe end, it is best to do only half at a time, or if a large top a quarter only.

The approved method of treating dining-table tops is to well body-in with French polish, after which thoroughly glass-paper down with fine paper, and then use the oil polish (see page 87).

Immediately after using a rubber, it should be kept in an air-tight tin canister, where it will always remain fresh and fit for use.

The Polishing Shop. — A few words as to the polishing shop may be acceptable to those who possess ample room and desire the best results.

First in order is the location and arrangement 53 of the finishing rooms. Preference is to be given to the upper rooms of a building for several reasons, among which may be named the securing of better light, greater freedom from dust, and superior ventilation.

A good light in this, as in many other arts, is a very important matter, and by a good light we mean all the light that can be obtained without the glare of the direct rays of the sun. Light from side windows is preferable to that from skylights for three reasons: (1) Skylights are very liable to leakage; (2) they are frequently, for greater or less periods, covered with snow in winter; (3) the rays of the sun transmitted by them in summer are frequently so powerful as to blister shellac or varnish.

Good ventilation is at all times of importance, and especially so in summer, both as tending to dry the varnish or shellac more evenly and rapidly, and as contributing to the comfort of the workmen. The latter consideration is of importance even as a matter of economy, as men in a room the atmosphere of which is pleasant and wholesome will feel better and accomplish more than they could do in the close and forbidding apartments in which they sometimes work.

Any suggestion in reference to freedom from dust, as a matter to be considered in locating rooms for this business, would seem to be entirely superfluous, as it is clear that there is hardly any 54 department of mechanical work which is so susceptible to injury from dust as the finishing of furniture, including varnishing and polishing.

Finishing rooms may be arranged in three departments. The first should include the room devoted to sand-papering and filling. These processes, much more than any other part of furniture polishing, produce dirt and dust, and it is plain that the room devoted to them should be so far isolated from the varnishing room as not to introduce into it these injurious elements.

Another room should be appropriated to the bodying-in, smoothing and rubbing-down processes. The third room is for spiriting and varnishing, or the application of the final coats of varnish, which is the most important of all the processes in finishing. It requires a very light and clean room, and a greater degree of heat than a general workroom. It should, as nearly as possible, be uniform, and kept up to *summer heat*; in no case ought the temperature to fall below fifty nor rise higher than eighty-five degrees Fahrenheit while the varnishing process is going on. Varnishing performed under these circumstances will be more thorough in result, have a brighter appearance and better polish, than if the drying is slow and under irregular temperature. For drying work, the best kind of heat is that from a stove or furnace. 55

Steam heat is not so good for two reasons: (1), it is too moist and soft, causing the work to sweat rather than to dry hard, and (2), the temperature of a room heated by steam is liable to considerable variation, and especially to becoming lower in the night. This *fire heat* is as necessary for the varnishing room in damp and cloudy weather in summer as it is in winter. At all seasons, and by night as well as by day, the heat should be as dry as possible, and kept uniformly up to summer heat, by whatever means this result is secured. Varnished work, after receiving the last coat, should be allowed to remain one day in the varnishing room. It may then be removed into the general workroom.

A remark may be proper here, viz., that there is sometimes a failure to secure the best and most permanent results from not allowing sufficient time for and between the several processes. An order is perhaps to be filled, or for some other reason the goods are "rushed through" at the cost of thoroughness and excellence of finish.

The following suggestion is made by way of caution in reference to the disposal of oily rags and waste made in the various processes of finishing. These articles are regarded as very dangerous, and are frequently the cause of much controversy between insurance companies and parties who are insured. The best way to dispose of this waste is to put it into the stove and 56 burn it as fast as it is produced. If this rule is strictly adhered to there will be no danger of fire from this source. All liquid stock should be kept in close cans or barrels, and as far from the fire as possible.

57

CHAPTER VIII.

ENAMELLING.

The process of enamelling in oil varnishes as applied to furniture must be understood as a smooth, glossy surface of various colours produced by bodies of paint and varnish skilfully rubbed down, and prepared in a peculiar way so as to produce a surface equal to French polish. Ornament can be added by gilding, etc., after the polished surface is finished.

We will begin with the white or light-tinted enamel. The same process must be pursued for any colour, the only difference being in the selection of the materials for the tint required to be produced.

It should be observed that enamelling requires the exercise of the greatest care, and will not bear hurrying. Each coat must be allowed sufficient time for the hardening, and the rubbing down must be patiently and gently done; heavy pressure will completely spoil the work.

Materials.—The materials used for the purpose above named are: white lead ground in 58 turpentine and the best white lead in oil; a clear, quick, and hard-drying varnish, such as the best copal, or the varnishes for enamel manufactured by Mr. W. Urquhart, 327, Edgware Road, W.; or white coburg and white enamel varnish, ground and lump pumice-stone, or putty-powder, great care being taken in the selection of the pumice-stone, as the slightest particle of grit will spoil the surface; and rotten-stone, used either with water or oil.

Tools.—The tools required are several flat wooden blocks, of various sizes and forms, suitable for inserting into corners and for mouldings—these must be covered with felt on the side you intend to use, the felt best adapted for the purpose being the white felt, from a quarter to half an inch in thickness, which can be obtained of Messrs. Thomas Wallis & Co., Holborn Circus, or at the woollen warehouses; two or three bosses (made similar to polish rubbers) of cotton-wool, and covered with silk (an old silk handkerchief makes capital coverings); wash or chamois leather, and a good sponge.

Mode of Operation.—If the wood is soft and porous it is best to commence with a coating of size and whiting applied in a warm state, which is allowed to dry; it is then rubbed down with 59 glass-paper, and two coats of common paint given, mixed in the usual way and of the same colour as you intend to finish with. In practice this is found to be best; after these two coats are thoroughly dry, mix the white-lead ground in turps, with only a sufficient quantity of varnish to bind it, thinning to a proper consistency with turps. It is as well to add a little of the ordinary white-lead ground in oil, as it helps to prevent cracking. Give the work four or five coats of this, and allow each coat to dry thoroughly. When it is hard and ready for rubbing down, commence with a soft piece of pumice-stone and water, and rub just sufficient to take off the roughness. Now use the felt-covered rubbers and ground pumice-stone, and cut it down, working in a circular manner. The greatest care is required to obtain a level surface free from scratches.

After the work is well rubbed down, if it should appear to be insufficiently filled up, or if scratched, give it two more coats, laid on very smoothly, and rub down as before. If properly done, it will be perfectly smooth and free from scratches. Wash it well down, and be careful to clean off all the loose pumice-stone. Then mix flake-white from the tube with either of the above-named varnishes, till it is of the consistency of cream. Give one coat of this, and when dry give it another, adding more varnish. Let this dry hard, the time taken for which will of course depend upon the drying qualities of the varnish; 60 some will polish in eight or nine days, but it is much the best to let it stand as long as you possibly can, as the harder it is, the brighter and more enduring will be the polish. When sufficiently hard, use the felt, and very finely-ground pumice-stone and water; with this cut down till it is perfectly smooth; then let it stand for a couple of days, to harden the surface.

Polishing.—In commencing to bring up a polish, first take rotten-stone, either in oil or water; use this with the felt rubber for a little while, then put some upon the surface of the silk-covered boss, and commence to rub very gently in circular strokes; continue this till there is a fine equal surface all over. The polish will begin to appear as you proceed, but it will be of a dull sort. Clean off: if the rotten-stone is in oil, clean off with dry flour; if in water, wash off with sponge and leather, taking care that you wash it perfectly clean and do not scratch.

You will now, after having washed your hands, use a clean damp chamois leather, holding it in the left hand, and using the right to polish with, keeping it clean by frequently drawing it over the damp leather. With the ball of the right hand press gently upon the work, and draw your hand sharply, forward or towards you; this will produce a bright polish, and every time you bring 61 your hand forward a sharp shrill sound will be heard similar to rubbing on glass. Continue this till the whole surface is one bright even polish. It will be some time before you will be able to do this perfectly, especially if the skin is dry or hard, as it is then liable to scratch the work. A smooth, soft skin will produce the best polish.

For the interior of houses, the "Albarine" enamel manufactured by the Yorkshire Varnish Company, of Ripon, is recommended. This article combines in itself a perfectly hard solid enamel of the purest possible colour; and for all interior decorations, where purity of colour and brilliancy of finish are desired, it is universally admitted to be the most perfect article of the kind hitherto introduced to the trade. It is applied in the same manner as ordinary varnish.

Another Process.—The preceding section describes the process of enamelling by oil varnishes, and the directions referring to the polishing will be found of value for the "polishing up" on painted imitations of woods or marbles. There is another process whereby an enamel can be produced upon furniture at a much cheaper rate than the preceding, and one too, perhaps, in which a polisher may feel more "at home." The work should first have a coating of size and whiting (well strained); this will act as a pore-filler. When dry, rub down with fine paper, after which use the felt-covered rubber and 62 powdered pumice-stone, to remove all the scratches caused by the glass-paper and to obtain a smooth and good surface. Then proceed to make a solution for the enamel: first procure two ounces of common isinglass from the druggist's, and thoroughly dissolve it in about a pint of boiling water; when dissolved, stir in two ounces and a-half of subnitrate of bismuth—this will be found to be about the right quantity for most woods, but it can be varied to suit the requirements. With this give the work one coat, boiling hot; apply it with a soft piece of Turkey sponge, or a broad camel's-hair brush, and when dry cut down with powdered pumice-stone; if a second coat is required, serve in precisely the same manner. Then proceed to polish in the ordinary way with white polish. After wetting the rubber, sprinkle a small quantity of the subnitrate of bismuth upon it; then put on the cover, and work in the usual manner; continue this till a sufficient body is obtained, and after allowing a sufficient time for the sinking and hardening it can be spirited off.

Enamelled furniture has had, comparatively speaking, rather a dull sale, but there is no class of furniture more susceptible of being made to please the fancy of the many than this. It can be made in any tint that may be required by the application of Judson's dyes, and the exercise of a little skill in the decoration will produce very pleasing effects. 63

Decorations.—The decorations are usually ornaments drawn in gold. A cut-out stencil pattern is generally used, and the surface brushed over with a camel's-hair pencil and japanner's gold size, which can be obtained at the artist's colourman's, or, if preferred, can be made by boiling 4 ozs. of linseed-oil with 1 oz. of gum anîme and a little vermilion. When the size is tacky, or nearly dry, gold powder or gold leaf is applied. The gold is gently pressed down with a piece of wadding, and when dry the surplus can be removed with a round camel's-hair tool. In all cases where gold has been fixed by this process it will bear washing without coming off, which is a great advantage.

CHAPTER IX.

AMERICAN POLISHING PROCESSES

The method of polishing furniture practised by the American manufacturers differs considerably from the French polishing processes adopted by manufacturers in most European countries. This difference, however, is mostly compulsory, and is attributable to the climate. The intense heat of summer and the extreme cold of winter will soon render a French polish useless, and as a consequence numerous experiments have been tried to obtain a polish for furniture that will resist heat or cold. The writer has extracted from two American cabinet-trade journals, *The Cabinet-maker* and *The Trade Bureau*, descriptions of the various processes now used in the States, which descriptions were evidently contributed by practical workmen. The following pages are not, strictly speaking, a mere reprint from the above-named journals, the articles having been carefully revised and re-written after having been practically tested; attention to them is, therefore, strongly recommended. 65

In these processes the work is first filled in with a "putty filler," and after the surface has been thoroughly cleaned it is ready for shellac or varnish. Second, a coating of shellac is next applied with a brush or a soft piece of Turkey sponge. This mixture is composed of two parts (by weight) of shellac to one of methylated spirits, but what is called "thin shellac" is composed of one part shellac to two of spirits. After the coating is laid on and allowed to dry, which it does very soon, it is rubbed carefully with fine flour glass-paper, or powdered pumice-stone—about four coats are usually given, each one rubbed down as directed. Third, when the surface has received a sufficient body, get a felt-covered rubber and apply rotten-stone and sweet oil in the same manner as you would clean brass; with this give the work a good rubbing, so as to produce a polish. Fourth, clean off with a rag and sweet oil, and rub dry; then take a soft rag with a few drops of spirit upon it, and vapour up to a fine polish.

With these few preliminary remarks, the following will be easily understood.

Use Of Fillers.—The cost of a putty filler consists chiefly in the time consumed in applying it. In the matter of walnut-filling much expense is saved in the processes of coating and rubbing if the pores of the wood be filled to the surface with a substance that will not shrink, and will harden 66 quickly. The time occupied in spreading and cleaning a thin or fatty mixture of filler, or a stiff and brittle putty made fresh every day, is about the same, and while the thin mixture will be subject to a great shrinkage, the putty filler will hold its own. It will thus be seen that a proper regard to the materials used in making fillers, and the consistency and freshness of the same, form an important element in the economy of filling.

A principal cause of poor filling is the use of thin material. By some a putty-knife is used, and the filling rubbed into the surfaces of mouldings with tow, while others use only the tow for all surfaces, mostly, however, in cases of dry filling. In the use of the wet filler, either with a knife or with tow, workmen are prone to spread it too thin because it requires less effort, but experience shows that the greatest care should always be taken to spread the putty stiff and thick, notwithstanding the complaints of workmen. In fact, this class of work does not bring into play so much muscle as to warrant complaints on account of it. Nor can there be any reasonable excuse for taking a longer time to spread a stiff filler than a thin filler.

Good results are not always obtained by the use of thick fillers, because the putty is spread too soon after the application of the first coat of oil, which liquid should be quite thin, and reduced either with benzine or turpentine, so that when 67 the putty is forced into the pores the oil already in them will have the effect of thinning it. As an illustration of the idea meant here to be conveyed, we will suppose a quantity of thick mud or peat dumped into a cavity containing water, and a similar quantity of the same material dumped into another cavity having no water; the one fills the bottom of the cavity solid, while the other becomes partly liquid at the bottom, and must of necessity shrink before it assumes the solidity of the former. Hence it appears that work to be filled should be oiled and allowed to stand some time before receiving the filler, or until the oil has been absorbed into the pores.

The preparatory coating should not be mixed so as to dry too quickly, nor allowed to stand too long before introducing the putty, for in this case the putty when forced along by the knife will not slip so easily as it should.

The cost of rubbing and sand-papering in the finishing process is very much lessened if the cleaning be thorough, and if all the corners and mouldings be scraped out, so that pieces of putty do not remain to work up into the first coat of shellac, or whatever finish may be used as a substitute for shellac.

Another important feature in hard filling is to let the work be well dried before applying the first coat of finish. One day is not sufficient for the proper drying of putty fillers, and if in consequence of insufficient drying a part of the 68 filling washes out, it is so much labour lost. As a safeguard against washing out, these fillers should be mixed with as much dryer or japan as the case warrants, for it frequently occurs that work must be finished, or go into finish, the day following the filling, whether it be dry or not.

By observing the main facts here alluded to, good filling may always be obtained, and at a cost not exceeding that of poor work.

For the light woods, including ash, chestnut, and oak, the filling is similar to that used in walnut, except the colouring material, which, of course, must be slight, or just enough to prevent the whiting and plaster from showing white in the pores. This colouring may consist of raw sienna, burnt sienna, or a trifle raw, or umber; one of these ingredients separate, or all three combined, mixed so as to please the fancy and suit the prevailing style. The colouring may be used with a dry filling, although a wet filling is more likely to give a smooth finish and greater satisfaction, and the colour of the filler can be seen better in the putty than in the dry powder.

Upon cheap work a filler should be used that requires the least amount of labour in its application. For this purpose liquid fillers, like japan, are suitable. If, however, a fine finish on fine goods is required, the putty compositions of various mixtures are the more appropriate. The secret of the process of filling consists in the mix69 ing of the compounds and the method of using them. A liquid filler or a japan simply spread over the work in one or two coats can hardly be called filling, yet this will serve the purpose very well for cheap furniture.

Thick compositions or putty fillers are composed of whiting and plaster, or similar powders having little or no colour. This material is mixed with oil, japan, and benzine, with a sufficient quantity of colouring matter to please the fancy. The value of these fillers is in proportion to their brittleness or "shortness," as it is termed, and, to give them this quality, plaster is used and as much benzine or turpentine as the mixture will bear without being too stiff or too hard to clean off. Sometimes a little dissolved shellac is used to produce "shortness." This desirable feature of a filler is best effected by mixing a small quantity of the material at a time. Many workmen mistakenly mix large batches at a time with a view of securing uniformity of colour, and this is one cause why such fillers work tough and produce a poor surface. An oil mixture soon becomes fatty and tough, and must be reduced in consistency when used, as it is apt when old to "drag" and leave the pores only partly filled. These fillers should be mixed fresh every day, and allowed to stiffen and solidify in the wood rather than out of it.

The surface of a pore is the largest part of it, and it is desirable to fill it to a level as nearly as 70 possible. This is done by using the filler thick or stiff.

Making Fillers.—In making "fillers," a quantity of the japan which is used in the ingredients can be made at one time, and used from as occasion may require. It is made in the following manner:

Japan of the Best Quality.—Put ¾ lb. gum shellac into 1 gall. linseed-oil; take ½ lb. each of litharge, burnt umber, and red-lead, also 6 oz. sugar of lead. Boil in the mixture of shellac and oil until all are dissolved; this will require about four hours. Remove from the fire, and stir in 1 gall. of spirits of turpentine, and the work is finished.

Fillings for Light Woods.—Take 5 lb. of whiting, 3 lb. calcined plaster (plaster of Paris), ½ gall. of raw linseed-oil, 1 qt. of spirits of turpentine, 1 qt. of brown japan, and a little French yellow to tinge the white. Mix well, and apply with a brush; rub it well with excelsior or tow, and clean off with rags. This thoroughly fills the pores of the wood and preserves its natural colour.

Another for Light Woods. — Take 10 lb. of whiting, 5 lb. of calcined plaster, 1 lb. of corn starch, 3 oz. calcined magnesia, 1 gall. of raw linseed-oil, ½ gall. spirits of turpentine, 1 qt. of brown japan, 2 oz. French yellow. Mix well, 71 and apply with brush; rub in well with excelsior or tow, and clean off with rags.

For Mahogany or Cherry Wood. — Take 5 lb. of whiting, 2 lb. of calcined plaster, 1½ oz. dry burnt sienna, 1 oz. Venetian red, 1 qt. of boiled linseed-oil, 1 pt. of spirits of turpentine, and 1 pt. of brown japan. Mix well, apply with brush, and rub well in with excelsior or tow. Clean off with rags dry.

For Oak Wood. — Take 5 lb. of whiting, 2 lb. calcined plaster, 1 oz. dry burnt sienna, ½ oz. of dry French yellow, 1 qt. raw linseed-oil, 1 pt. benzine spirits, and ½ pt. white shellac. Mix well, apply with brush, rub in with excelsior or tow, and clean off with rags.

For Rosewood. — Take 6 lb. of fine whiting, 2 lb. of calcined plaster, 1 lb. of rose-pink, 2 oz. of Venetian red, ½ lb. of Vandyke brown, ½ lb. of Brandon red, 1 gall. of boiled linseed-oil, ½ gall. of spirits of turpentine, 1 qt. of black japan. Mix well together, apply with brush, rub well in with tow, and clean off with rags.

For Black Walnut (1). — For medium and cheap work. Take 10 lb. of whiting, 3 lb. dry burnt umber, 4 lb. of Vandyke brown, 3 lb. of calcined plaster, ½ lb. of Venetian red, 1 gall. of boiled linseed-oil, ½ gall. of spirits of turpentine, 1 qt. of black japan. Mix well and apply with brush; rub well with excelsior or tow, and clean off with rags.

For Black Walnut (2). — An improved filling, 72 producing a fine imitation of wax finish, may be effected by taking 5 lb. of whiting, with 1 lb. of calcined plaster, 6 oz. of calcined magnesia, 1 oz. of dry burnt umber, 1 oz. of French yellow to tinge the white. Add 1 qt. of raw linseed-oil, 1 qt. of benzine spirits, ½ pt. of very thin white shellac. Mix well, and apply with a brush; rub well in, and clean off with rags.

An Oil-Colour for Black Walnut (3), to be used only on first-class and custom work. — Take 3 lb. of burnt umber ground in oil, 1 lb. of burnt sienna ground in oil, 1 qt. of spirits of turpentine, 1 pt. of brown japan. Mix well and apply with a brush. Sand-paper well; clean off with tow and rags. This gives a beautiful chocolate colour to the wood.

Numerous compositions are in the market for filling the pores of wood, and in this connection particular attention has been given to walnut, for the reason that this wood is used in large quantities in the furniture industry, and is nearly, if not quite, as porous as any other of the woods used.

A variety of walnut fillings have been recommended to the trade in order to meet the demand consequent upon the different grades of finish and the method of obtaining the finish, so that it would be difficult to pronounce as to the superiority of any one filling for general purposes. In treating this subject, attention should be given to the necessities for the use of filling, so that each one 73 may determine for himself the kind of composition best adapted for the work in hand, and the best method of applying it.

Finishing.—Having described the methods of making and applying the "fillings," we will now describe the mode of finishing, and begin with the "dead-oil finish." We can remember when a satisfactory oil-finish was produced either with a good quality of japan or a fair quality of spirits. These materials are recommended to be used by inexperienced workmen and those not familiar with the mixing of the various grades of japan and varnish with oil, turpentine, benzine, etc. This method of oil-finish, too, is scarcely inferior to the shellac or spirit-varnish method, and it is cheaper. When the best finish is desired, a sufficient number of coats to fill the pores of wood to a level are required, and then the whole surface should be subjected to the rubbing process. The use of these fillers provides an oil-finish in a simplified form for those who are not aware of the difference between hard and soft gum compositions as a base for rubbing. In fact, the rubbing process constitutes a fine oil-finish, and requires a hard gum, whether it be of japan, varnish, or shellac.

The use of varnish or its substitute as a filler and finish is more frequent than the use of shellac, and for cheap work it is equally good. 74 The surface produced by a hard gum composition must be smooth and dead, or but slightly glossed, so as to admit of the pores being filled full or to a level. It may be added that a coat or any number of coats of the composition referred to above is substantially a filling, and the quality of finish depends upon the number of coats, together with the amount of rubbing applied.

Thus far we have simply called attention to the best quality of oil-finish and the manner of producing it. Possibly three-fourths of all wood-finishing, particularly walnut-finishing, is several degrees below the best quality. In fact, oil-finish may imply only one coat of any composition that will dry, while two coats may be regarded as fair, and three coats a very good quality of finish. For the class of finish not rubbed down with pumice-stone and water, oil-varnish would be out of place on account of its gloss; hence shellac, being in composition similar to japan, is the better material, because of its dull appearance or lack of gloss as compared with shellac.

In addition to the liquid fillers already mentioned, there is a putty or powder filling used for cross-grained woods, or such woods as have a deep pore. This filling is forced into the wood previous to the application of the other finishing compounds, with the use of which it in no way interferes. On the contrary, it economises the use of the liquid fillers, and, while constituting a part of an oil-finish, is also a finish wholly independent of 75 the other methods mentioned — that is to say, the same results can be obtained by the use of either one, although the putty or powder filling is attended with greater expense both as to time and material. The hard filling is generally used on walnut, ash, and all coarse-grained woods.

With regard to oil-finishes, viz., spirit-varnish or oil-varnish, shellac is thought by many to be the best for fine work; but others think differently. We may say of shellac that it will finish up into any degree of polish, and while it will not retain a French polish long in this climate, it will replenish easier and cheaper than any other finish, and continue to improve under each application. For a common finish, however, oil preparation is as good as shellac, and even for a fine finish it is only second to shellac, if made of a hard gum. On common finish, too, the oil will wear better than shellac in stock or on storage, so far as preserving its freshness is concerned.

The cost of oil-finish is governed chiefly by the amount of labour expended on it. A suite of walnut furniture can be well rubbed with sand-paper in two hours, or even less; while two weeks could be profitably employed in rubbing another suite with pumice and water.

Black Walnut Finishing.—The fashionable finish for black walnut work, particularly chamber sets, is what is known to the trade as the "dead-oil finish." It is admired, perhaps, because it has a gloss, rather than a shine of the varnish stamp. There is no more labour required upon it than upon a bright finish, but the process of manipulation is different, and harder to the fingers.

It should be premised that the walnut work of the day bears upon its surface, to a greater or less extent, raised panels covered with French burl veneer. And upon this fact largely depends the beauty of the production. And the endeavour is to so finish the article that there shall be a contrast between the panel and the groundwork on which it is placed. In other words, the former should be of a light colour, while the latter is of a darker shade. In that view the palest shellac should be used on the panels, and darker pieces, liver coloured, etc., on the body of the work. The darker grades of shellac are the cheaper, and will answer for the bulk of the work, but the clearest only for the panels.

In commencing to finish a job direct from the cabinet-maker's hand, rough and innocent of sand-paper, first cover the panels with a coat of shellac to prevent the oil in the filling from colouring them dark. Next, cover the body of the work with a wood filling composed of whiting and plaster of Paris, mixed with japan, benzine, and raw linseed-oil, or the lubricating oil made from petroleum; the whole covered with umber, to which, in the rare cases when a reddish shade is wanted, Venetian red is also added. This filling is then rubbed off with cloths, and by this process tends to close up the grain of the wood and produce an even surface. More or less time should be allowed after each of the several steps in the finishing process for the work to dry and harden, though much less is required in working with shellac than with varnishes composed of turpentine, oil, and gums. But the time that should be allowed is often lessened by the desire to get the work through as soon as possible, so that no standard can be set up as to the number of hours required between each of the several processes. It would be well if twelve hours intervened, but if work to which ten days could well be devoted must be hurried through in three, obviously the processes must follow each other in a corresponding haste.

A coating of shellac is then given the whole work, light on the panels and dark on the body work, and when it has dried and hardened, which it does very soon, it may be rubbed down. This process of "rubbing down" should be done evenly and carefully, so as not to rub through the shellac at any point, and be done with the finer grades of sand-paper for the cheaper class of work, particularly at first, but at a later period of the process, and for the better class of articles in all cases, hair-cloth should be used, the material for the "rubbing down" being pumice-stone moistened with raw linseed-oil for the best work, and the 78 lubricating oil, before mentioned, for cheaper work, or the covered parts of the better grades. This rubbing down involves labour, wear of fingers and finger-nails, and is carried on with an ordinary bit of hair-cloth, the smooth surface next the wood, and not made in any particular shape, but as a wad, ball, or otherwise. In the corners and crevices where the hair-cloth will not enter it will be necessary to use sand-paper of the finest grades, and worn pieces only.

Three coats of shellac are put on, followed each time by this rubbing-down process, each one giving the work a smoother feeling and a more perfect appearance. Afterwards, to complete the whole, a coating of japan thinned with benzine is applied, which gives to the work a clean appearance and the dead glossy finish.

There is this objection to the above style of finish, that the japan catches all the dust which touches it, and holds it permanently, so that many of the best workmen will not have work finished in this way for their own private houses, preferring the brighter look given by shellac and varnish without rubbing down the last coat, believing that the work can be kept much cleaner.

Finishing Veneered Panels, etc.—The large oval panels of desks, etc., covered with French veneer, are generally taken out and finished by themselves. The process is similar 79 to that above given, with successive coats of shellac and varnish, and the oil and pumice-stone rubbing down; but the final part of this latter process is a rubbing down with rotten-stone; then the merest trifle of sweet-oil is applied all over the surface and wiped off. (See Rosewood, etc., farther on.)

For Light Woods (Dead Finish).—Apply two or three coats of white shellac; rub down with pumice and raw linseed-oil, and clean off well with rags; use varnish-polish on the panels.

Another.—Finish as in the previous recipe. For a flowing coat of varnish-finish apply one flowing coat of light amber varnish. If a varnish-polish is desired, apply three coats of Zanzibar polishing varnish. Rub down and polish, and the result will be a splendid finish.

Mahogany or Cherry Wood.—For shellac *dead finish* apply two coats of yellow shellac. Rub down with pumice and raw linseed-oil. If a varnish-finish is desired, apply a flowing coat of light amber varnish or shellac thus rubbed. The panels should receive two coats of Zanzibar polishing varnish.

Oak.—For a *dead finish* give three coats of shellac, two-thirds of white and one-third of yellow, mixed. Rub down with pumice and raw linseed-oil. For a cheap varnish-finish give one flowing coat of light amber varnish in the shellac, rubbed as directed. Varnish-polish the panels.

Rosewood, Coromandel, or Kingwood (a Bright 80 Finish).—Apply two thin coats of shellac, sand-papering each coat; then apply three or four coats of Zanzibar polishing varnish, laying it on thin, and giving it sufficient time to dry thoroughly. When it is perfectly hard, rub down with pumice and water. Polish with rotten-stone to a fine lustre, clean up with sweet-oil, and vapour up the oil with a damp alcohol rag. The result is a splendid mirror-like polish. This is the method employed in polishing pianofortes in America.

Walnut.—For a cheap finish, apply one coat of yellow shellac. When dry, sand-paper down. Apply with brush; rub in well; clean off with rags. This gives a very fair finish.

For a medium *dead finish* apply two or three coats of yellow shellac. When dry, rub down with pumice and raw linseed-oil; clean up well; varnish-polish the panels.

For *finish*. Before using the above filling, give the work one coat of white shellac. When dry, sand-paper down, and apply the above filling. Give two coats of white shellac; rub down with pumice and raw linseed-oil; clean up well with brown japan and spirits of turpentine, mixed. Wipe off. This is a good imitation of wax-finish; it is waterproof, and will not spot as wax-finish does. The panels are to be varnished-polished. This is to be used with the improved filling No. 2.

For *finish*. Apply three coats of yellow shellac; rub down with pumice and raw linseed81 oil; clean off well. Varnish-polish the panels. Use this with the oil colour No. 3.

Finishing Cheap Work.—*With One Coat of Varnish.*—Give the work a coat of boiled linseed-oil; immediately sprinkle dry whiting upon it, and rub it well in with tow all over the surface. The whiting absorbs the oil and completely fills the pores of the wood. For black walnut add a little dry burnt umber. For mahogany or cherry add a little Venetian red, according to the colour of the wood. The application can be made to turned work while in motion in the lathe. Clean off well with rags. The work can then be finished with a single coat of varnish, and for cheap work makes a very good finish.

For varnishing large surfaces, a two-inch oval varnish brush is to be used first to lay out the varnish, and then a two-inch flat badger flowing-brush for a softener. The latter lays down moats and bubbles left by the large brush. A perfectly smooth glass-like surface is thus obtained. When not in use, these tools should be put into a pot containing raw linseed-oil and spirits of turpentine. This keeps them in a better working condition than if they are kept in varnish, making them clean and soft. Standing in varnish they congeal and become hard as the spirit evaporates from the varnish. For shellacing a large surface use a two-inch bristle brush; for small work, such as 82 carvings and mouldings, use a one-and-a-half inch flat brush. These brushes when not in use should be taken from the various pots and deposited in an earthen pot sufficiently large to hold all the shellac brushes used in the shop. Put in enough of raw linseed-oil and thin shellac to cover the bristles of the brushes. Kept in this manner, they will remain clean and elastic, and will wear much longer.

Wax Finishing.—Take ½ gall. of turpentine, 1½ lb. yellow beeswax, 1 lb. white beeswax, ½ lb. white rosin. Pulverise the rosin, and shave the wax into fine shavings. Put the whole into the turpentine, and dissolve it cold. If dissolved by a fire-heat, the vitality of the wax is destroyed. When it is thoroughly dissolved, mix well and apply with a stiff brush. Rub well in, and clean off with rags. When dry, it is ready for shellac or varnish as may be desired.

A Varnish Polish.—Take 10 oz. gum shellac, 1 oz. gum sandarach, 1 drachm Venice turpentine, 1 gall. alcohol. Put the mixture into a jug for a day or two, shaking occasionally. When dissolved it is ready for use. Apply a few coats. Polish by rubbing smooth.

For the commonest kind of work in black walnut a very cheap polish can be made in the following manner: Take 1 gall. of turpentine, 2 lb. pulverised asphaltum, 1 qt. boiled linseed-oil, 2 oz. Venetian red. Put the mixture in a warm place and shake occasionally. When it is 83 dissolved, strain and apply to the wood with a stiff brush. Rub well with cloth when dry. Then take 1 pt. of thin shellac, ½ pt. boiled linseed-oil. Shake it well before using. Apply with cloth, rubbing briskly, and you will have a fine polish.

With Copal or Zanzibar Varnish.—As a substitute for filling, the wood may receive one coat of native coal-oil, thinned with benzine-spirits; then apply one coat of shellac, and follow with varnish, as desired. The time is not far distant when manufacturers must and will use varnish for the finishing of all kinds of furniture on account of the high price of shellac. Furniture finished in the last-named method may be rubbed with either water or oil. Water has a tendency to harden varnish, while oil softens it. If water is used there will be a saving of oil and rags. In the other case shellac, when rubbed with oil, should be cleaned with japan. This removes the greasy and cloudy appearance which is left after the rubbing with oil, and the work will have a clean, dry, and brighter appearance than otherwise.

We suggest another idea for finishing black walnut for a cheap or a medium class of work. In the first place, fill the pores of the wood, and apply one thin coat of shellac to hold the filling in the pores of the wood. Let this stand one day; sand-paper down with fine paper, then with a brush apply a coat of coach japan. Rub well, 84 and clean off with rags. Let this stand one day to dry, then, with some sand-paper that has been used before, take off the moats from the japan. Go over the whole surface with a soft rag saturated with japan; wipe and clean off carefully, and the job is finished. This, though a cheap finish, is a good one for this class of work.

We give one more method of finishing black walnut, that is, with boiled linseed-oil only, and there is no other way of obtaining a genuine oil-finish. Sand-paper the wood down smoothly; apply a coat of boiled linseed-oil over the whole surface; sand-paper well, and clean up dry with rags; let it stand one day to dry, then apply one more coat of oil; rub well in with rags, but do not use sand-paper on this coat. Apply three, four, or more coats in the same way. When the work has received the last coat of oil and is dry, sand-paper down with old paper. Then clean up with the best coach japan with rags, and let the work stand one day to dry. The panels are to be varnish-polished the same as other wood. The work is then finished, and ready for the warerooms.

This method takes a longer time than finishing with either varnish or shellac; but the cost is less both for materials and for labour, the workman being able to go over a greater surface in the same time. The work will stand longer, and the method gives a rich and close finish, bringing out 85 the figure and rich colour of the wood better than in any other method of finishing. It does not cost so much as shellac finish; it only requires a little more time for drying between the coats of oil. In finishing in varnish or shellac, to get the body or surface for polishing three or four coats are frequently applied, which is liable to produce a dull cloudy appearance. For this reason, and having in view the high and increasing price of stock, it seems to us that this really superior method of finishing in oil must take the place of shellac and varnish-finish in good work.

Polishing Varnish.—This is certainly a tedious process, and considered by many a matter of difficulty. The following is the mode of procedure: Put two ounces of powdered tripoli into an earthen pot or basin, with water sufficient to cover it; then, with a piece of fine flannel four times doubled, laid over a piece of cork rubber, proceed to polish your varnish, always wetting it well with the tripoli and water. You will know when the process is complete by wiping a part of the work with a sponge and observing whether there is a fair and even gloss. Clean off with a bit of mutton suet and fine flour. Be careful not to rub the work too hard, or longer than is necessary to make the face perfectly smooth and even. Some workmen polish with rotten-stone, others with putty-powder, and others with common whiting and water; but tripoli, we think, will be found to answer best.

An American Polish Reviver.—Take of olive-oil 1 lb., of rectified oil of amber 1 lb., spirits of turpentine 1 lb., oil of lavender 1 oz., tincture of alkanet-root ½ oz. Saturate a piece of cotton batting with this polish, and apply it to the wood; then, with soft and dry cotton rags, rub well and wipe off dry. This will make old furniture in private dwellings, or that which has been shop-worn in warerooms, look as well as when first finished. The articles should be put into a jar or jug, well mixed, and afterwards kept tightly corked.

This is a valuable recipe, and is not known, the writer believes, outside of his practice.

CHAPTER X.

MISCELLANEOUS RECIPES.

Oil Polish.—One quart of cold-drawn linseed-oil to be simmered (not boiled) for ten minutes, and strained through flannel; then add one-eighth part of spirits of turpentine: to be applied daily with soft linen rags, and rubbed off lightly; each time the oil is applied the surface should be previously washed with cold water, so as to remove any dirt or dust. This method of polishing is particularly useful for dining-table tops; it will in about six weeks produce a polish so durable as to resist boiling water or hot dishes, and be like a mirror for brilliancy.

Wax Polish.—Eight ounces of beeswax, 2 oz. of resin, and ½ oz. of Venetian turpentine, to be melted over a slow fire; the mass, when quite melted, is poured into a sufficiently large stone-ware pot, and while it is still warm 6 oz. of rectified turpentine are stirred in. After the lapse of twenty-four hours the mass will have assumed the consistency of soft butter, and is ready for use. A small portion of the polish is taken up with a woollen rag and rubbed over the surface of the work—at first gently, then more strongly. When the polish is uniformly laid on, the surface is once more rubbed lightly and quickly with a fresh clean rag to produce a gloss.

Waterproof French Polish.—Take 2 oz. gum benjamin, ½ oz. gum sandarach, ½ oz. gum anîme, 1½ oz. gum benzoin, and 1 pt. alcohol. Mix in a closely-stoppered bottle, and put in a warm place till the gums are well dissolved. Then strain off, and add ¼ gill of poppy-oil. Shake well together, and it is ready for use.

A Varnish for Musical Instruments.—Take one gallon of alcohol, 1 lb. gum sandarach, ½ lb. gum mastic, 2 lbs. best white resin, 3 lbs. gum benzoin; cut the gums cold. When they are thoroughly dissolved, strain the mixture through fine muslin, and bottle for use; keep the bottle tightly corked. This is a beautiful varnish for violins and other musical instruments of wood, and for fancy articles, such as those of inlaid work. It is also well adapted for panel-work, and all kinds of cabinet furniture. There is required only one flowing coat, and it produces a very fine mirror-like surface. Apply this varnish 89 with a flat camel's-hair or sable brush. In an hour after application the surface is perfectly dry.

French Varnish for Cabinet-work.—Take of shellac 1½ oz. gum mastic and gum sandarach, of each ½ oz., spirit of wine by weight 20 oz. The gums to be first dissolved in the spirit, and lastly the shellac. This may be best effected by means of the water-bath. Place a loosely-corked bottle containing the mixture in a vessel of warm water of a temperature below the boiling point, and let it remain until the gums are dissolved. Should evaporation take place, an equal quantity to the spirit of wine so lost must be replaced till the mixture settles, then pour off the clear liquid for use, leaving the impurities behind; but do not filter it. Greater hardness may be given to the varnish by increasing the quantity of shellac, which may be done to the amount of one-twelfth of the lac to eleven-twelfths of spirit. But in this latter proportion the varnish loses its transparency in some degree, and must be laid on in very small quantities at a time.

Mastic Varnish.—Mastic should be dissolved in oil of turpentine, in close glass vessels, by means of a gentle heat. This varnish is extensively used in transparencies, etc. 90

Cabinet-maker's Varnish.—Take 5 lbs. very pale gum shellac, 7 oz. gum mastic, 1 gallon alcohol. Dissolve in a cold atmosphere with frequent stirring.

Amber Varnish.—This is a most difficult varnish to make. It is usually prepared by roasting the amber and adding hot linseed-oil, after which turpentine can be mixed if required. But for a small quantity, dissolve the broken amber, without heat, in the smallest possible quantity of chloroform or pure benzine. Heat the linseed-oil, remove it from the fire, and pour in the amber solution, stirring all the time. Then add the turpentine. If not quite clear, heat again, using the utmost caution.

Colourless Varnish with Copal.—To prepare this varnish the copal must be picked; each piece is broken, and a drop of rosemary-oil poured on it. Those pieces which, on contact with the oil, become soft are the ones used. The pieces being selected, they are ground and passed through a sieve, being reduced to a fine powder. It is then placed in a glass, and a corresponding volume of rosemary-oil poured over it; the mixture is then stirred for a few minutes until it is transformed into a thick liquor. It is then left to rest for two hours, when a few drops of rectified 91 alcohol are added, and intimately mixed. Repeat the operation until the varnish is of a sufficient consistency; leave the rest for a few days, and decant the clear. This varnish can be applied to wood and metals (*Journal of Applied Chemistry*).

Seedlac Varnish.—Wash 3 oz. of seedlac in several waters; dry it and powder it coarsely. Dissolve it in one pint of rectified spirits of wine; submit it to gentle heat, shaking it as often as convenient, until it appears dissolved. Pour off the clear part, and strain the remainder.

Patent Varnish for Wood or Canvas.—Take 1 gallon spirits of turpentine, 2¼ lbs. asphaltum. Put them into an iron kettle on a stove, and dissolve the gum by heat. When it is dissolved and a little cool, add 1 pint copal varnish and 1 pint boiled linseed-oil.

When entirely cool it is ready for use. For a perfect black add a little lamp-black.

Copal Varnish.—Dissolve the copal, broken in pieces, in linseed-oil, by digestion, the heat being almost sufficient to boil the oil. The oil should be made drying by the addition of quick-lime. This makes a beautiful transparent varnish. It should be diluted with oil of turpentine; a 92 very small quantity of copal, in proportion to the oil, will be found sufficient.

Carriage Varnish.—Take 19 oz. gum sandarach, 9½ oz. orange shellac, 12½ oz. white resin, 18 oz. turpentine, 5 pints alcohol. Dissolve and strain. Use for the internal parts of carriages and similar purposes. This varnish dries in ten minutes.

Transparent Varnish.—Take 1 gallon alcohol, 2 lbs. gum sandarach, ½ 1b. gum mastic. Place them in a tin can. Cork tight and shake frequently, placing the can in a warm place. When dissolved it is ready for use.

Crystal Varnish for Maps, etc.—Mix together 1 oz. Canada balsam and 2 oz. spirits of turpentine. Before applying this varnish to a drawing or a painting in water-colours the paper should be placed on a stretcher, sized with a thin solution of isinglass in water, and dried. Apply the varnish with a soft camel's-hair brush.

A Black Varnish.—Mix a small quantity of gas-black with the brown hard varnish pre93 viously mentioned. The black can be obtained by boiling a pot over a gas-burner, so that it almost touches the burner, when a fine jet-black will form at the bottom, which remove and mix with the varnish, and apply with a brush.

A Black Polish can be made in the same way: after wetting the rubber, just touch it with the black. Place the linen cover over, touch it with oil, and it is ready for work.

Varnish for Iron.—Take 2 lbs. pulverised gum asphaltum, ¼ lb. gum benzoin, 1 gallon spirits of turpentine. To make this varnish quickly, keep in a warm place, and shake often till it is dissolved. Shade to suit with finely-ground ivory-black. Apply with a brush. This varnish should be used on iron-work exposed to the weather. It is also well adapted for inside work, such as iron furniture, where a handsome polish is desired.

Varnish for Tools.—Take 2 oz. tallow, 1 oz. resin; melt together, and strain while hot to remove the specks which are in the resin. Apply a slight coat on the tools with a brush, and it will keep off the rust for any length of time.

To Make Labels Adhere to a Polished Surface.—Brush the back of a label over with thin varnish or polish, and press down with a soft rag; this must be done quickly, as the polish soon becomes dry. This is the way labels are put on pianofortes, and also the paper imitation of fancy woods on polished pine-work.

How to Remove French Polish or Varnish from Old Work.—Cleaning off old work for re-polishing or varnishing is usually found difficult, and to occupy much time if only the scraper and glass-paper be used. It can be easily accomplished in a very short time by washing the surface with liquid ammonia, applied with a piece of rag; the polish will peel off like a skin, and leave the wood quite bare. In carvings or turned work, after applying the ammonia, use a hard brush to remove the varnish. Unadulterated spirits of wine used in a tepid state will answer the same purpose.

Colouring for Carcase Work.—In the best class of cabinet-work all the inside work—such as carcase backs, shelves, etc.—is made of good materials, such as wainscot, soft mahogany, Havannah cedar, or American walnut; but for second-class work, pine or white deal is used instead, and coloured.

The colouring matter used should match with the exterior wood. For mahogany take ½ lb. of ground yellow ochre to a quart of water, and add about a tablespoonful of Venetian red—a very small quantity of red in proportion to the yellow is sufficient for mahogany—and a piece of glue about the size of a walnut; the whole to be well stirred and boiled. Brush over while hot, and immediately rub off with soft shavings or a sponge. For the antique hues of old wainscot mix equal parts of burnt umber and brown ochre. For new oak, bird's-eye maple, birch, satin-wood, or any similar light yellowish woods, whiting or white-lead, tinted with orange chrome, or by yellow ochre and a little size. For walnut, brown umber, glue size, and water; or by burnt umber very moderately modified with yellow ochre. For rosewood, Venetian red tinted with lamp-black. For ebony, ivory-black; but for the common ebonised work lamp-black is generally used.

When the colouring is dry, it should be rubbed down with a piece of worn fine glass-paper, and polished with beeswax rubbed on a very hard brush—a worn-out scrubbing-brush is as good as anything—or it can be well rubbed with Dutch rush. In polishing always rub the way of the grain. The cheap work seldom gets more than a coat of colour rubbed off with shavings.

Cheap but Valuable Stain for the Sap of Black Walnut.—Take 1 gallon of strong 96 vinegar, 1 lb. dry burnt umber, ½ lb. fine rose-pink, ½ lb. dry burnt Vandyke brown. Put them into a jug and mix them well; let the mixture stand one day, and it will then be ready for use. Apply this stain to the sap with a piece of fine sponge; it will dry in half an hour. The whole piece is then ready for the filling process. When completed, the stained part cannot be detected even by those who have performed the work. This recipe is of value, as by it wood of poor quality and mostly of sap can be used with good effect.

Polish for Removing Stains, etc., from Furniture (American).—Take ½ pint alcohol, ¼ oz. pulverised resin, ¼ oz. gum shellac, ½ pint boiled linseed-oil. Shake the mixture well, and apply it with a sponge, brush, or cotton flannel, rubbing well after the application.

Walnut Stain to be used on Pine and White-wood.—Take 1 gallon of very thin sized shellac; add 1 lb. of dry burnt umber, 1 lb. of dry burnt sienna, and ¼ lb. of lamp-black. Put these articles into a jug, and shake frequently until they are mixed. Apply one coat with a brush. When the work is dry, sand-paper down with fine paper, and apply one coat of shellac or cheap varnish. It will then be a good imitation 97 of solid walnut, and will be adapted for the backboards of mirror-frames, for the backside and inside of case-work, and for similar work.

Rosewood Stain.—Take 1 lb. of logwood chips, ½ lb. of red-sanders, ½ gallon of water. Boil over a fire until the full strength is obtained. Apply the mixture, while hot, to the wood with a brush. Use one or two coats to obtain a strong red colour. Then take 1 gallon of spirits of turpentine and 2 lb. of asphaltum. Dissolve in an iron kettle on a stove, stirring constantly. Apply with a brush over the red stain, to imitate rosewood. To make a perfect black, add a little lamp-black. The addition of a small quantity of varnish with the turpentine will improve it. This stain applied to birchwood gives as good an imitation of rosewood as on black walnut, the shade on the birch being a little brighter.

Rosewood Stain for Cane Work, etc.—Take 1 gallon alcohol, 1 lb. red-sanders, 1 lb. dragon's blood, 1 lb. extract logwood, ½ lb. gum shellac. Put the mixture into a jug, and steep well till it obtains its full strength. Then strain, and it will be ready for use. Apply with brush, giving one, two, or more coats, according to the depth of colour desired. Then give one or more coats of varnish. This stain is suitable for use 98 on cane, willow, or reed work, and produces a good imitation of rosewood.

French Polish Reviver.—This recipe will be found a valuable one. If the work is sweated and dirty, make it tolerably wet, and let it stand a few minutes; then rub off and polish with a soft rag. It is important that the ingredients should be mixed in a bottle in the order as given: Vinegar, 1 gill; methylated spirit, 1 gill; linseed-oil, ½ pint; butter of antimony (poison), 1 oz. Raw linseed-oil, moderately thinned with turpentine or spirits of wine, will also make a good reviver. Old furniture, or furniture that has been warehoused for a long time, should be washed with soda and warm water previous to applying the reviver.

Morocco Leather Reviver.—The coverings of chairs or sofas in morocco, roan, or skiver can be much improved by this reviver. If old and greasy, wash with sour milk first. The reviver should be applied with a piece of wadding, and wiped one way only, as in glazing. The colour can be matched by adding red-sanders. Methylated spirit, ½ pint; gum benzoin, 2 oz.; shellac, ½ oz. Mix, and shake up occasionally until dissolved.99

Hair-cloth Reviver.—Mix equal parts of marrow-oil (neats-foot), ox-gall. and ivory-black, to be well rubbed with a cloth. This composition forms a valuable renovator for old hair-cloth.

To Remove Grease Stains from Silks, Damasks, Cloth, etc.—Pour over the stain a small quantity of benzoline spirit, and it will soon disappear without leaving the least mark behind. The most delicate colours can be so treated without fear of injury. For paint stains chloroform is very efficacious.

To Remove Ink Stains from White Marble.—Make a little chloride of lime into a paste with water, and rub it into the stains, and let it remain a few hours; then wash off with soap and water.

CHAPTER XI.

MATERIALS USED.

Alkanet-root (botanical name, *Anchusa tinctoria*).—This plant is a native of the Levant, but it is much cultivated in the south of France and in Germany. The root is the only part used by French polishers to obtain a rich quiet red; the colouring is chiefly contained in the bark or outer covering, and is easily obtained by soaking the root in spirits or linseed-oil. The plant itself is a small herbaceous perennial, and grows to about a foot in height, with lance-shaped leaves and purple flowers, and with a long woody root with a deep red bark.

Madder-root (*Rubia tinctoria*).—This plant is indigenous to the Levant; but it is much cultivated in Southern Europe, and also in India. Its uses are for dyeing and staining; it can be procured in a powdered state, and imparts its red colour when soaked in water or spirits. This is a creeping plant with a slender stem; almost quadrangular, the leaves grow four in a bunch; flowers small, fruit yellow, berry double, one being abortive. The roots are dug up when the plant has attained the age of two or three years; they are of a long cylindrical shape, about the thickness of a quill, and of a red-brownish colour, and when powdered are a bright Turkish-red. Extracts of madder are mostly obtained by treating the root with boiling water, collecting the precipitates which separate on cooling, mixing them with gum or starch, and adding acetate of alumina or iron. This is in fact a mixture of colouring matter and a mordant.

Red-sanders (*Pterocarpus santalinus*).—The tree from which this wood is obtained is a lofty one, and is to be found in many parts of India, especially about Madras. It yields a dye of a bright garnet-red colour, and is used by French polishers for dyeing polishes, varnishes, revivers, etc.

Logwood (*Hæmatoxylon campeachianum*).—This is a moderate-sized tree with a very contorted trunk and branches, which are beset with sharp thorns, and blooms with a yellow flower. It is a native of Central America and the West Indies. This valuable dyewood is imported in logs; the heart-wood is the most valuable, which 102 is cut up into chips or ground to powder for the use of dyers by large powerful mills constructed especially for the purpose. Logwood, when boiled in water, easily imparts its red colour. If a few drops of acetic acid (vinegar) is added, a bright red is produced; and when a little alum is added for a mordant, it forms red ink. If an alkali, such as soda or potash, is used instead of an acid, the colour changes to a dark blue or purple, and with a little management every shade of these colours can be obtained. Logwood put into polish or varnish also imparts its red colour.

Fustic (*Maclura tinctoria*).—This tree is a native of the West Indies, and imparts a yellow dye. Great quantities are used for dyeing linens, etc. The fustic is a large and handsome evergreen, and is imported in long sticks.

Turmeric (*Curcuma longa*).—Turmeric is a stemless plant, with palmated tuberous roots and smooth lance-shaped leaves. It is imported from the East Indies and China. The root is the part which affords the yellow powder for dyeing. It is also a condiment, and is largely used in Indian curry-powder. Paper stained with turmeric is used by chemists as a test for alkalies, and it is also used in making Dutch, pink, and gold-coloured varnishes.103

Indigo (*Indigofera tinctoria*).—Indigo is a shrub which grows from two to three feet in height, and is cut down just as it begins to flower. It is cultivated in almost all the countries situated in the tropics. The dye substance is prepared from the stems and leaves, and is largely used in calico-printing.

Persian Berries (*Rhamnus infectorius*).—These berries are the produce of a shrub of a species of buckthorn common in Persia, whence they derive their name; but large quantities are also imported into England from Turkey and the south of France. The berries are gathered in an unripe state, and furnish a yellow dye.

Nut-galls.—These are found upon the young twigs of the Turkish dwarf oak (*Quercus infectoria*), and are produced by the puncture of an insect called Cynips. The supply is principally from Turkey and Aleppo. Nut-galls contain a large quantity of tannin and gallic acid, and are extensively used in dyeing.

Catechu.—This is obtained from the East Indies, and is the extract of the *Acacia catechu*, a thorny tree. The wood is cut up into chips similar to logwood, and after boiling and evapora104 tion the liquor assumes the consistency of tar; but when cold it hardens, and is formed into small squares. It is extensively used by tanners in place of oak bark.

Thus.—Thus is the resin which exudes from the spruce-fir, and is used by some polishers in the making of polishes and varnishes.

Sandarach is the produce of the *Thuya articulata* of Barbary. It occurs in small pale yellow scales, slightly acid, and is soluble in alcohol; it is used in both polishes and varnishes.

Mastic exudes from the mastic-tree (*Pistacia lentiscus*), and is principally obtained from Chios, in the Grecian Archipelago. It runs freely when an incision is made in the body of the tree, but not otherwise. It occurs in the form of nearly colourless and transparent tears of a faint smell, and is soluble in alcohol as well as oil of turpentine, forming a rapidly-drying but alterable varnish, which becomes brittle and dark-coloured by age.

Benzoin.—This is the produce of the American tree *Laurus benzoin*, and also of the *Styrax benzoin* 105 of Sumatra, which is called "gum benjamin"; it is used in polishes and varnishes, and as a cosmetic, and is also burnt as incense in Catholic churches.

Copal is one of the most valuable of gums, and is furnished by many countries in the districts of Africa explored by Mr. H. M. Stanley, the discoverer of Livingstone. Copal is found in a fossil state in very large quantities. The natives collect the gum by searching in the sandy soil, mostly in the hilly districts, the country being almost barren, with no large tree except the Adansonia, and occasionally a few thorny bushes.

The gum is dug out of the earth by the copal gatherers at various depths, from two or three to ten or more feet, in a manner resembling gold-digging; and great excitement appears when a good amount is discovered. The gum is found in various shapes and sizes, resembling a hen's egg, a flat cake, a child's head, etc. There are three kinds, yellow, red, and whitish; and the first furnishes the best varnish and fetches the highest price from the dealers. Many of the natives assert that the copal still grows on different trees, and that it acquires its excellent qualities as a resin by dropping off and sinking several feet into the soil, whereby it is cleansed, and obtains, after a lapse of many years, its hardness, inflammability, and transparency.106

Dragon's Blood is the juice of certain tropical plants of a red colour, especially of the tree *Pterocarpus draco*. After the juice is extracted, it is reduced to a powder by evaporation. It is used for darkening mahogany, colouring varnishes or polishes, etc., and for staining marble. Chemists also use it in preparing tinctures and tooth powders.

Shellac—or, more properly, *gum-lac*—is a resinous substance obtained from the Bihar-tree, and also from the *Ficus Indica*, or Banyan-tree. It exudes when the branches are pierced by an insect called the *Coccus ficus*. The twigs encrusted with the resin in its natural state is called Stick-lac. When the resin is broken off the twigs, powdered, and rubbed with water, a good deal of the red colouring matter is dissolved, and the granular resin left is called seed-lac; and when melted, strained, and spread into thin plates it is called shellac, and is prepared in various ways and known by the names of button, garnet, liver, orange, ruby, thread, etc., and is used for many purposes in the arts. Shellac forms the principal ingredient for polishes and spirit varnishes. Red sealing-wax is composed of shellac, Venice turpentine, and vermilion red; for the black sealing-wax ivory-black is used instead of the vermilion. Shellac is soluble in alcohol, and in many acids and alkalies. Lac-dye is the red 107 colour from the stick-lac dissolved by water and evaporated to dryness. The dye, however, is principally from the shrivelled-up body of the insect of the Stick-lac.

Shellac is produced in the largest quantity and the best quality in Bengal, Assam, and Burmah. The chief seat of manufacture is Calcutta, where the native manufacturers are accused of adulterating it with resin to a considerable extent. The best customers are Great

Britain and the United States, though the demand in the Italian markets appears to be on the increase.

Amber is a yellow, semi-transparent, fossil resin; hard but brittle, and easily cut with a knife; tasteless, and without smell, except when pounded or heated, and then it emits a fragrant odour. It has considerable lustre; becomes highly electric by friction; and will burn with a yellow flame. It is found in nodules of various sizes in alluvial soils, or on the seashore in many places, particularly on the shores of the Baltic. Amber is much employed for ornamental purposes, and is also used in the manufacture of amber-varnish. It will not dissolve in alcohol, but yields to the concentrated action of sulphuric acid, which will dissolve all resins except caramba wax.

Pumice-stone.—This well-known light and 108 spongy volcanic substance is extensively quarried in the small islands that lie off the coast of Sicily. Its porosity and smooth-cutting properties render it of great value to painters and polishers for levelling down first coatings. Ground pumice-stone is the best for cutting down bodies of polish or varnish that are more advanced towards completion. The best way to get a surface to a piece of lump pumice-stone is to rub it down on a flat York stone, or, better still, an old tile that has been well baked. Pumice-stone should not be allowed to stand in water; it causes the grain to contract and to harden, thereby deteriorating its cutting properties.

Linseed-oil.—This valuable oil is obtained by pressure from the seed of the flax plant (*Linum usitatissimum*). Linseed contains on an average about 33 per cent. of oil, though the amount varies materially, the percentage obtained fluctuating considerably, not being alike on any two successive days. This is partly due to the varying richness of the seed, and partly to the manner in which it is manipulated in extracting the oil, it being a very easy matter to lose a considerable percentage of the oil by a lack of skill in any of the processes, though they all seem so simple.

The first thing done with the seed from which the oil is to be extracted is to pass it through a 109 screen, to cleanse it from foreign substances. The seed is received in bags containing from three to four bushels, and pockets containing one-sixth of that amount. Having been screened it is passed through a mill, whose large iron-rollers, three in number, grind it to a coarse meal. Thence it is carried to what are known as the "mullers," which are two large stones, about eight feet in diameter and eighteen inches thick, weighing six tons each, standing on their edges, and rolling around on a stone bed. About five bushels of the meal are placed in the mullers, and about eight quarts of hot water are added. The meal is afterwards carried by machinery to the heaters, iron pans holding about a bushel each. These are heated to an even temperature by steam, and are partly filled with the meal, which for seven minutes is submitted to the heat, being carefully stirred in order that all parts may become evenly heated. At the end of that time the meal is placed in bags, which in turn are placed in hydraulic presses, iron plates being placed between the bags. Pressure is applied for about eight minutes, until, as is supposed, all the oil is pressed out, leaving a hard cake, known to the trade as oil-cake, or linseed-cake.

The product of these various processes is known as "raw" oil, a considerable portion of which is sold without further labour being expended upon it. There is, however, a demand for "boiled" oil, for certain purposes where greater drying pro110 perties are needed. To supply this want oil is placed in large kettles, holding from five hundred to one thousand gallons, where it is heated to a temperature of about 500 degrees, being stirred continually. This process, when large kettles are used, requires nearly the entire day. While the boiling process is going on, oxide of manganese is added, which helps to give the boiled oil better drying properties. A considerable portion of the oil is bleached, for the use of manufacturers of white paints.

Venice Turpentine.—This is obtained from the larch, and is said to be contained in peculiar sacs in the upper part of the stem, and to be obtained by puncturing them. It is a ropy liquid, colourless or brownish green, having a somewhat unpleasant odour and bitter taste.

Oil of Turpentine is the most plentiful and useful of oils. It is obtained in America from a species of pine very plentiful in the Carolinas, Georgia, and Alabama, known as the long-leaved pine (*pinus Australis*), and found only where the original forest has not been removed.

Methylated Spirits.—The methylated spirit of commerce usually consists of the ordinary mixed grain, or "plain" spirit, as produced by the large distillers in London and elsewhere, with 111 which are blended, by simply mixing in various proportions, one part vegetable naphtha and three parts spirits of wine. The mixing takes place in presence of a revenue officer, and the spirits so "methylated" are allowed to be used duty free. The revenue authorities consider the admixture of naphtha, having so pungent and disagreeable a smell, a sufficient security against its sale and consumption as a beverage. No process has yet been discovered of getting rid of this odour. It is illegal for druggists to use it in the preparation of medicinal tinctures, unless they are for external use.

PRINTED BY WILLIAM CLOWES AND SONS, LIMITED, LONDON AND BECCLES.

Crosby Lockwood & Son's

LIST OF WORKS

ON

TRADES AND MANUFACTURES, THE INDUSTRIAL ARTS, CHEMICAL MANUFACTURES, COUNTING HOUSE WORK, Etc.

A Complete Catalogue of NEW and STANDARD BOOKS relating to CIVIL, MECHANICAL, MARINE and ELECTRICAL ENGINEERING; MINING, METALLURGY, and COLLIERY WORKING; ARCHITECTURE and BUILDING; AGRICULTURE and ESTATE MANAGEMENT, etc. Post Free on Application.

7, STATIONERS' HALL COURT, LONDON, E.C.,
AND
121a, VICTORIA STREET, WESTMINSTER, S.W.
1910.

LIST OF WORKS

ON

TRADES and MANUFACTURES, THE INDUSTRIAL ARTS, Etc.

ACETYLENE, LIGHTING BY. Generators, Burners, and Electric Furnaces. By William E. Gibbs, M.E. With 66 Illustrations. Crown 8vo, cloth 7/6

AIR GAS LIGHTING SYSTEMS. See Petrol Gas.

ALCOHOL (INDUSTRIAL): ITS MANUFACTURE AND USES. A Practical Treatise based on Dr. Max Maercker's "Introduction to Distillation," as revised by Drs. Delbruck and Lange. By J. K. Brachvogel. 500 pages, 105 engravings *Net* **16/6**

The Industrial Value of Tax-Free Alcohol and what it means to Agricultural Interests — Summary of the Processes in Spirit Manufacture — Starch, How Formed, its Characteristics, and the Changes it Undergoes — Enzymes or Ferments — Products of Fermentation — Starchy and Sacchariferous Raw Materials — Preparation of the Malt — Steaming the Raw Material — The Mashing Process — Fermenting the Mash — Preparation of Artificial Yeast in the Distillery — Fermentation in Practice — Distillation and Rectification — Arrangement of the Distillery — The Spent Wash — Denaturing of Alcohol — Alcohol for the Production of Power, Heating and Illumination — Statistics.

ALKALI TRADE MANUAL. Including the Manufacture of Sulphuric Acid, Sulphate of Soda, and Bleaching Powder. By John Lomas, Alkali Manufacturer. With 232 Illustrations. Super-royal 8vo, cloth. £1 10s.

BLOWPIPE IN CHEMISTRY, MINERALOGY, Etc. Containing all known Methods of Anhydrous Analysis, many Working Examples, and Instructions for Making Apparatus. By Lieut.-Colonel W. A. Ross, R.A., F.G.S. Second Edition. Crown 8vo, cloth 5/0

BOOT AND SHOE MAKING, including Measurement, Last-fitting, Cutting-out, Closing and Making, with a Description of the most Approved Machinery employed. By J. B. Leno. Crown 8vo, cloth 2/0

BRASS FOUNDER'S MANUAL. Modelling, Pattern Making, Moulding, Turning, &c. By W. Graham. Crown 8vo, cloth 2/0

BREAD & BISCUIT BAKER'S & SUGAR-BOILER'S ASSISTANT. Including a large variety of Modern Recipes. By Robert Wells. Fifth Edition. Crown 8vo, cloth 1/0

"A large number of wrinkles for the ordinary cook, as well as the baker." — *Saturday Review.*

BREAKFAST DISHES. For every Morning of Three Months. By Miss Allen (Mrs. A. Macaire). Author of "Savouries and Sweets," &c. Twenty-third Edition. F'cap 8vo. Sewed 1/0
Or, quarter bound, fancy boards 1/6

BREWERS, HANDY BOOK FOR. Being a Practical Guide to the Art of Brewing and Malting. Embracing the Conclusions of Modern Research which bear upon the Practice of Brewing. By H. E. Wright, M.A. Third Edition. Thoroughly Revised and Enlarged. Large Crown 8vo, 578 pp., cloth *Net* **12/6**

Barley, Malting and Malt — Water for Brewing — Hops and Sugars — The Brewing Room — Chemistry as Applied to Brewing — The Laboratory — Mashing, Sparging, and Boiling — Ferments in General — Fermentation With Commercial Yeast — Culture from a Single Cell with Yeasts — Treatment of Beer — The Brewery and Plants.

"We have great pleasure in recommending this handy Book." — *The Brewers' Guardian.*

CALCULATOR, NUMBER, WEIGHT AND FRACTIONAL. Containing upwards of 250,000 Separate Calculations, showing at a Glance the Value at 422 Different Rates, ranging from 1/128th of a Penny to 20s. each, or per cwt., and £20 per ton, of any number of articles consecutively, from 1 to 470. Any number of cwts., qrs., and lbs., from 1 cwt. to 470 cwts. Any number of tons, cwts., qrs., and lbs., from 1 to 1,000 tons. By William Chadwick, Public Accountant. Fourth Edition, Revised and Improved. 8vo, strongly bound 18/0

"It is as easy of reference for any answer or any number of answers as a dictionary. For making up accounts or estimates the book must prove invaluable to all who have any considerable quantity of calculations involving price and measure in any combination to do." —*Engineer.*

"The most perfect work of the kind yet prepared." —*Glasgow Herald.*

CEMENTS, PASTES, GLUES, AND GUMS. A Guide to the Manufacture and Application of Agglutinants for Workshop, Laboratory, or Office Use. With 900 Recipes and Formulæ. By H. C. Standage, Crown 8vo, cloth 2/0

"As a revelation of what are considered trade secrets, this book will arouse an amount of curiosity among the large number of industries it touches." —*Daily Chronicle.*

CHEMISTRY FOR ARMY AND MATRICULATION CANDIDATES. By Geoffrey Martin, B.Sc., Ph.D. Crown 8vo, cloth. With numerous Illustrations *Net* **2/0**

Preparation and Use of Apparatus — Preparation and Properties of Certain Gases and Liquids — Simple Quantitative Experiments — Analytical Operations — Solubility — Water Crystallisation — Neutralisation of Acids by Bases, and Preparation of Simple Salts — Volumetric Analysis — Chemical Equivalents — Observation of Reaction — Melting and Boiling Points — Symbols and Atomic Weights of the Elements — Weights and Measures — Hints on Regulating Work in Practical Chemistry Classes.

CLOCKS, WATCHES, & BELLS for PUBLIC PURPOSES. By Edmund Beckett, Lord Grimthorpe, LL.D., K.C., F.R.A.S. Eighth Edition, with new List of Great Bells and an Appendix on Weathercocks. Crown 8vo, cloth 4/6; cloth boards, 5/6

"The only modern treatise on clock-making." —*Horological Journal.*

COACH-BUILDING. A Practical Treatise, Historical and Descriptive. By J. W. Burgess. Crown 8vo, cloth 2/6

COKE—MODERN COKING PRACTICE. Including the Analysis of Materials and Products. A handbook for those engaged or interested in Coke Manufacture with recovery of By-Products. By T.

H. Byrom, F.I.C., F.C.S., Mem. Soc. of Chem. Industry, Chief Chemist to the Wigan Coal and Iron Company. For fifteen years Lecturer at the Wigan Technical College. Author of "The Physics and Chemistry of Mining"; and J. E. Christopher, Mem. Soc. of Chem. Industry, Sub-manager of the Semet Solvay Coking Plant of the Wigan Coal and Iron Company. Lecturer on Coke Manufacture at the Wigan Technical College. 168 pages, with numerous illustrations. Demy 8vo, cloth. [*Just Published Net*] **8/6**

"The authors have succeeded in treating the subject in a clear and compact way, giving an easily comprehensible review of the different processes." — *Mining Journal.*

"The book will be eagerly read, and the authors may be assured that their labour will be appreciated. We anticipate that the book will be a success; at any rate it possesses the necessary merit." — *Science and Art of Mining.*

COMMERCIAL CORRESPONDENT, FOREIGN. Being Aids to Commercial Correspondence in Five Languages—English, French, German, Italian, and Spanish. By Conrad E. Baker. Third Edition, Carefully Revised Throughout. Crown 8vo, cloth 4/6

"Whoever wishes to correspond in all the languages mentioned by Mr. Baker cannot do better than study this work, the materials of which are excellent and conveniently arranged. They consist not of entire specimen letters, but—what are far more useful—short passages, sentences, or phrases expressing the same general idea in various forms." — *Athenæum.*

CONFECTIONER, MODERN FLOUR. Containing a large Collection of Recipes for Cheap Cakes, Biscuits, &c. With remarks on the Ingredients Used in their Manufacture. By R. Wells 1/0

CONFECTIONERY, ORNAMENTAL. A Guide for Bakers, Confectioners and Pastrycooks; including a variety of Modern Recipes, and Remarks on Decorative and Coloured Work. With 129 Original Designs. By Robert Wells. Crown 8vo, cloth 5/0

"A valuable work, practical, and should be in the hands of every baker and confectioner. The illustrative designs are worth treble the amount charged for the work." — *Bakers' Times.*

COTTON MANUFACTURE. A Manual of Practical Instruction of the Processes of Opening, Carding, Combing, Drawing, Doubling and Spinning, Methods of Dyeing, &c. For the Use of Operatives, Overlookers, and Manufacturers. By J. Lister. 8vo, cloth 7/6

DANGEROUS GOODS. Their Sources and Properties, Modes of Storage and Transport. With Notes and Comments on Accidents arising therefrom. For the Use of Government and Railway Officials, Steamship Owners, &c. By H. J. Phillips. Crown 8vo, cloth 9/0

DENTISTRY (MECHANICAL). A Practical Treatise on the Construction of the Various Kinds of Artificial Dentures. By C. Hunter. Crown 8vo, cloth 3/0

DISCOUNT GUIDE. Comprising several Series of Tables for the Use of Merchants, Manufacturers, Ironmongers, and Others, by which may be ascertained the Exact Profit arising from any mode of using Discounts, either in the Purchase or Sale of Goods, and the method of either Altering a Rate of Discount, or Advancing a Price, so as to produce, by one operation, a sum that will realise any required Profit after allowing one or more Discounts: to which are added Tables of Profit or Advance from 1¼ to 90 per cent., Tables of Discount from 1¼ to 98¾ per cent., and Tables of Commission, &c., from ⅛ to 10 per cent. By Henry Harben, Accountant. New Edition, Corrected. Demy 8vo, half-bound £1 5s.

"A book such as this can only be appreciated by business men, to whom the saving of time means saving of money. The work must prove of great value to merchants, manufacturers, and general traders." —*British Trade Journal*.

DRYING MACHINERY AND PRACTICE. A Handbook on the Theory and Practice of Drying and Desiccating, with Classified Description of Installations, Machinery, and Apparatus, including also a Glossary of Technical Terms and Bibliography. By Thomas G. Marlow, Grinding, Drying, and Separating Machinery Specialist. Medium 8vo. About 250 pages, with 150 Illustrations [*In the Press, price about* **12/6** *net.*]

ELECTRICITY IN FACTORIES AND WORKSHOPS: ITS COST AND CONVENIENCE. A Handbook for Power Producers and Power Users. By A. P. Haslam, M.I.E.E. 328 pages, with numerous illustrations. Large crown, 8vo, cloth *Net* **7/6**

ELECTRO-METALLURGY. A Practical Treatise. By Alexander Watt. Tenth Edition, enlarged and revised. Including the most Recent Processes. Crown 8vo, cloth 3/6

ELECTRO-PLATING. A Practical Handbook on the Deposition of Copper, Silver, Nickel, Gold, Aluminium, Brass, Platinum, &c., &c. By J. W. Urquhart, C.E. Fifth Edition, Revised. Crown 8vo, cloth 5/0

ELECTRO-PLATING & ELECTRO-REFINING OF METALS Being a new edition of Alexander Watt's "Electro-Deposition." Revised and Rewritten by A. Philip, B.Sc., Principal Assistant to the Admiralty Chemist. Crown 8vo, cloth *Net* **12/6**

PART I. ELECTRO-PLATING — Preliminary Considerations — Primary and Secondary Batteries — Thermopiles — Dynamos — Cost of Electrical Installations of Small Output for Electro-Plating — Historical Review of Electro Deposition — Electro Deposition of Copper — Deposition of Gold by Simple Immersion — Electro Deposition of Gold — Various Gilding Operations — Mercury Gilding — Electro Deposition of Silver — Imitation Antique Silver — Electro Deposition of Nickel, Tin, Iron and Zinc, Various Metals and Alloys — Recovery of Gold and Silver from Wash Solutions — Mechanical Operations connected with Electro Deposition — Materials Used in Electro Deposition.
PART II. ELECTRO METALLURGY — Electro Metallurgy of Copper — Cost of Electrolytic Copper Refining — Current Density as a Factor in Profits — Some Important Details in Electrolytic Copper Refineries — Electrolytic Gold and Silver Bullion Refining — Electrolytic Treatment of Tin — Electrolytic Refining of Lead — Electrolytic Production of Aluminium and Electrolytic Refining of Nickel — Electro Galvanising.

"Eminently a book for the practical worker in electro-deposition." —*Engineer.*

ELECTRO-TYPING. The Reproduction and Multiplication of Printing Surfaces and Works of Art by the Electro-Deposition of Metals. By J. W. Urquhart, C.E. Crown 8vo, cloth 5/0

ENGINEERING CHEMISTRY. A Practical Treatise for the Use of Analytical Chemists, Engineers, Iron Masters, Iron Founders, Students and others. Comprising Methods of Analysis and Valuation of the Principal Materials used in Engineering Work, with numerous Analyses, Examples and Suggestions. By H. Phillips. Third Edition, Revised. Crown 8vo, 420 pp., with Illustrations, cloth *Net* **10/6**

EXPLOSIVES, MODERN, A HANDBOOK ON. A Practical Treatise on the Manufacture and Use of Dynamite, Gun-Cotton, Nitro-Glycerine and other Explosive Compounds, including Collodion-Cotton. With Chapters on Explosives in Practical Application. By M. Eissler, M.E. Second Edition, Enlarged. Crown 8vo, cloth 12/6

"A veritable mine of information on the subject of explosives employed for military, mining, and blasting purposes." —*Army and Navy Gazette.*

EXPLOSIVES: NITRO-EXPLOSIVES. The Properties, Manufacture, and Analysis of Nitrated Substances, including the Fulminates, Smokeless Powders, and Celluloid. By P. G. Sanford, F.I.C., F.C.S., Public Analyst to the Borough of Penzance. Second Edition, enlarged. With Illustrations. Demy 8vo, cloth *Net* **10/6**

Nitro-Glycerine — Nitro-Cellulose, etc. — Dynamite — Nitro-Benzol, Roburite, Bellite, Picric Acid, etc. — The Fulminates — Smokeless Powders in General — Analysis of Explosives — Firing Point, Heat Tests, Determination of Relative Strength, etc.

"One of the very few text-books in which can be found just what is wanted. Mr. Sanford goes steadily through the whole list of explosives commonly used, he names any given explosive, and tells of what it is composed and how it is manufactured. The book is excellent." —*Engineer.*

FACTORY ACCOUNTS: THEIR PRINCIPLES AND PRACTICE. A Handbook for Accountants and Manufacturers, with Appendices on the Nomenclature of Machine Details, the Income Tax Acts, the Rating of Factories, Fire and Boiler Insurance, the Factory and Workshop Acts, etc., including a Glossary of Terms and a large number of Specimen Rulings. By Emile Garcke and J. M. Fells. Fifth Edition, Revised and Enlarged. Demy 8vo, cloth 7/6

"A very interesting description of the requirements of Factory Accounts.... The principle of assimilating the Factory Accounts to the general commercial books is one which we thoroughly agree with." —*Accountants' Journal.*

FLOUR MANUFACTURE. A Treatise on Milling Science and Practice. By Friedrich Kick, Imperial Regierungsrath, Professor of Mechanical Technology in the Imperial German Polytechnic Institute, Prague. Translated from the Second Enlarged and Revised Edition. By H. H. P. Powles, A.M.Inst.C.E. 400 pp., with 28 Folding Plates, and 167 Woodcuts. Royal 8vo, cloth £1 5s.

"This invaluable work is the standard authority on the science of milling." —*The Miller.*

FRENCH POLISHING AND ENAMELLING. Including numerous Recipes for making Polishes, Varnishes, Glaze, Lacquers, Revivers, &c. By R. Bitmead. Crown 8vo, cloth 1/6

GAS ENGINEER'S POCKET-BOOK. Comprising Tables, Notes and Memoranda relating to the Manufacture, Distribution and Use of Coal Gas and the Construction of Gas Works. By H. O'Connor, A.M.Inst.C.E. Third Edition. Revised. Crown 8vo, leather. *Net* **10/6**

GENERAL CONSTRUCTING MEMORANDA — General Mathematical Tables — Unloading Materials and Storage — Retort House — Condensers — Boilers, Engines, Pumps, and Exhausters — Scrubbers and Washers — Purifiers — Gasholder Tanks — Gasholders — Workshop Notes — MANUFACTURING — Storing Materials — Retort House (Working) — Condensing Gas — Exhausters, etc. — Washing and Scrubbing — Purification — Gasholders (Care of) — Distributing Gas — Testing — Enriching Processes — Product Works — Supplementary.

"The book contains a vast amount of information." —*Gas World.*

GAS ENGINEERING. See Producer Gas Practice and Industrial Gas Engineering.

GAS FITTING. A Practical Handbook. By John Black. Revised Edition. With 130 Illustrations. Crown 8vo, cloth 2/6

GAS LIGHTING. See Acetylene.

GAS LIGHTING FOR COUNTRY HOUSES. See Petrol Air Gas.

GAS MANUFACTURE, CHEMISTRY OF. A Practical Manual for the use of Gas Engineers, Gas Managers and Students. By Harold M. Royle, Chief Chemical Assistant at the Beckton Gas Works. Demy 8vo, cloth, 340 pages, with numerous Illustrations and Coloured Plate. *Net* **12/6**

Preparation of Standard Solutions — Analysis of Coals — Description of Various Types of Furnaces — Products of Carbonisation at Various Temperatures — Analysis of Crude Gas — Analysis of Lime — Analysis of Ammoniacal Liquor — Analytical Valuation of Oxide of Iron — Estimation of Naphthalin — Analysis of Fire-Bricks and Fire-Clay — Art of Photometry — Carburetted Water Gas — Appendix containing Statutory and Official Regulations for Testing Gas. Valuable Excerpts from Various important papers on Gas Chemistry, Useful Tables, Memoranda, etc.

GAS WORKS. Their Construction and Arrangement, and the Manufacture and Distribution of Coal Gas. By S. Hughes, C.E. Ninth Edition. Revised by H. O'Connor, A.M.Inst.C.E. Crown 8vo 6/0

GOLD WORKING. JEWELLER'S ASSISTANT for Masters and Workmen, Compiled from the Experience of Thirty Years' Workshop Practice. By G. E. Gee. Crown 8vo 7/6

GOLDSMITH'S HANDBOOK. Alloying, Melting, Reducing, Colouring, Collecting, and Refining. Manipulation, Recovery of Waste, Chemical and Physical Properties; Solders, Enamels, and other useful Rules and Recipes, &c. By G. E. Gee, Sixth Edition. Crown 8vo, cloth 3/0

GOLDSMITH'S AND SILVERSMITH'S COMPLETE HANDBOOK. By G. E. Gee. Crown 8vo, half bound 7/0

HALL-MARKING OF JEWELLERY. Comprising an account of all the different Assay Towns of the United Kingdom, with the Stamps at present employed; also the Laws relating to the Standards and Hall-marks at the various Assay Offices. By G. E. Gee. Crown 8vo 3/0

HANDYBOOKS FOR HANDICRAFTS. By Paul N. Hasluck. See page 16.

HOROLOGY, MODERN, IN THEORY AND PRACTICE. Translated from the French of Claudius Saunier, ex-Director of the School of Horology at Macon, by Julien Tripplin, F.R.A.S., Besançon Watch Manufacturer, and Edward Rigg, M.A., Assayer in the Royal Mint. With Seventy-eight Woodcuts and Twenty-two Coloured Copper Plates. Second Edition. Super-royal 8vo, £2 2s. cloth; half-calf £2 10s.

"There is no horological work in the English language at all to be compared to this production of M. Saunier's for clearness and completeness. It is alike good as a guide for the student and as a reference for the experienced horologist and skilled workman." — *Horological Journal.*

ILLUMINATING AND MISSAL PAINTING ON PAPER AND VELLUM. A Practical Treatise on Manuscript Work, Testimonials, and Herald Painting, with Chapters on Lettering and Writing, and on Mediæval Burnished Gold. With two Coloured Plates. By Philip Whithard (First-class Diploma for Illumination and Herald Painting, Printing Trades Exhibition, 1906). 156 pages. Crown 8vo, cloth *Net* **4/0**

INTEREST CALCULATOR. Containing Tables at 1, 1½, 2, 2½, 3, 3½, 3¾, 4, 4½, 4¾ and 5 per cent. By A. M. Campbell, Author of "The Concise Calendar." Crown 8vo, cloth *Net* **2/6**

IRON AND METAL TRADES' COMPANION. For Expeditiously ascertaining the Value of any Goods bought or sold by Weight, from 1s. per cwt. to 112s. per cwt., and from one farthing per pound to one shilling per pound. By Thomas Downie. Strongly bound in leather, 396 pp. **9/0**

"A most useful set of tables. Nothing like them before existed." — *Building News.*

IRON-PLATE WEIGHT TABLES. For Iron Shipbuilders, Engineers and Iron Merchants. Containing the Calculated Weights of upwards of 150,000 different sizes of Iron Plates, from 1 ft. by 6 ins. by ¼ in. to 10 ft. by 5 ft. by 1 in. Worked out on the basis of 40 lbs. to the square foot of iron of 1 in. in thickness. By H. Burlinson and W. H. Simpson. 4to, half bound   £1 5s.

LABOUR CONTRACTS. A Popular Handbook on the Law of Contracts or Works and Services. By David Gibbons. Fourth Edition, with Appendix of Statutes by T. F. Uttley; Solicitor. F'cap. 8vo, cloth 3/6

LAUNDRY MANAGEMENT. A Handbook for use in Private and Public Laundries. Cr. 8vo, cloth 2/0

LAW FOR MANUFACTURERS, EMPLOYERS AND OTHERS, ETC. See "Every Man's Own Lawyer." A Handy-book of the Principles of Law and Equity. By a Barrister. Forty-seventh (1910) Edition, including the Legislation of 1909. 830 pp. Large crown 8vo, cloth [*Just Published.*] *Net* **6/8**

SUMMARY OF CONTENTS: Landlord and Tenant — Vendors and Purchasers — Contracts and Agreements — Conveyances and Mortgages — Joint-stock Companies — Partnership — Shipping Law — Dealings with Money — Suretiship — Cheques, Bills and Notes — Bills of Sale — Bankruptcy — Masters, Servants and Workmen — Insurance: Life, Accident, etc. — Copyright, Patents. Trade Marks — Husband and Wife, Divorce — Infancy, Custody of Children — Trustees and Executors — Taxes and Death Duties — Clergymen, Doctors, and Lawyers — Parliamentary Elections — Local Government — Libel and Slander — Nuisances — Criminal Law — Game Laws, Gaming, Innkeepers — Forms of Wills, Agreements, Notices, etc.

"A useful and concise epitome of the law." — *Law Magazine.*

"A complete digest of the most useful facts which constitute English law." — *Globe.*

"A dictionary of legal facts well put together. The book is a very useful one." — *Spectator.*

LEATHER MANUFACTURE. A Practical Handbook of Tanning, Currying, and Chrome Leather Dressing. By A. Watt. Fifth Edition, Revised and Enlarged. 8vo, cloth *Net* **12/6**

Chemical Theory of the Tanning Process — The Skin — Hides and Skins — Tannin or Tannic Acid — Gallic Acid — Gallic Fermentation — Tanning Materials — Estimation of Tannin — Preliminary Operations — Depilation or Unhairing Skins and Hides — Deliming or Bating — Tanning Butts for Sole Leather — Tanning Processes — Tanning by Pressure — Quick Tanning — Harness Leather Tanning — American Tanning — Hemlock Tanning — Tanning by Electricity — Chemical Tanning — Miscellaneous Processes — Cost of American Tanning — Manufacture of Light Leathers — Dyeing Leather — Manufacture of White Leather — Chrome Leather Manufacture — Box Calf Manufacture — Chamois or Oil Leather Manufacture — Currying — Machinery Employed in Leather Manufacture — Embossing Leather — Fellmongering — Parchment, Vellum, and Shagreen — Gut Dressing — Glue Boiling — Utilisation of Tanner's Waste.

"A sound, comprehensive treatise on tanning and its accessories." — *Chemical Review.*

LEATHER MANUFACTURE. PRACTICAL TANNING: A Handbook of Modern Processes, Receipts and Suggestions for the Treatment of Hides, Skins, and Pelts of every description, including various Patents relating to Tanning, with specifications. By Louis A. Flemming, American Tanner. Second Edition, in great part re-written, thoroughly revised, and much enlarged. Illustrated by six full-page Plates. Medium 8vo, cloth, 630 pages [*Just published.*] Net **28/0**

MAGNETOS FOR AUTOMOBILISTS, HOW MADE AND HOW USED. A Handbook of Practical Instruction in the Manufacture and Adaptation of the Magneto to the needs of the Motorist. By S. R. Bottone, late of the Collegio del Carmine, Turin, Author of "The Dynamo," "Ignition Devices," &c. Second Edition, enlarged. With 52 Illustrations. Crown 8vo, cloth Net **2/0**

MARBLE AND MARBLE WORKING. A Handbook for Architects, Sculptors, Marble Quarry Owners and Workers, and all engaged in the Building and Decorative Industries. Containing numerous Illustrations and thirteen Coloured Plates. By W. G. Renwick, Author of "The Marble Industry," "The Working of Marble for Decorative Purposes," etc. 240 pages. Medium 8vo, cloth 15/0

The Chemistry of Marble — Its Geological Formation — A short Classification of Marbles — Antiquity of the Marble Industry —

Ancient Quarries and Methods of Working — Modern Quarries and Quarrying Methods — Machinery used in Quarrying — European and American Systems compared — Marble as Building Material — Uses of Marble other than for Building Purposes-Sources of Production: Italian, French, Belgian, and Greek Marbles, etc. — Marbles of the United Kingdom and British Colonies — Continental Marble Working — Marble Working Machinery — Marble Working in the United States — American Machinery Described and Compared — Marble Working: A British industry — Marble Substitutes and Imitations — Practical Points for the Consideration of Architects — Hints on the Selection of Marble — List of Marbles in Ordinary Use, with Descriptive Notes and Instances of their Application.

MENSURATION AND GAUGING. A POCKET-BOOK containing Tables, Rules, and Memoranda for Revenue Officers, Brewers, Spirit Merchants, &c. By J. B. Mant. Second Edition. 18mo, leather. 4/0

"Should be in the hands of every practical brewer." —*Brewers' Journal.*

METRIC TABLES, A SERIES OF. In which the British Standard Measures and Weights are compared with those of the Metric System at present in Use on the Continent. By C. H. Dowling, C.E. 8vo, cloth 10/6

"Mr. Dowling's tables are well put together as a ready-reckoner for the conversion of one system into the other." —*Athenæum.*

METROLOGY, MODERN. A Manual of the Metrical Units and Systems of the present Century. With an Appendix containing a proposed English System. By Lowis d'A. Jackson, A.M.Inst.C.E., Author of "Aid to Survey Practice," etc. Large crown 8vo, cloth 12/6

"We recommend the work to all interested in the practical reform of our weights and measures." —*Nature.*

MOTOR CAR, THE. A Practical Manual for the use of Students and Motor Car Owners, with notes on the Internal Combustion Engine and its fuel. By Robert W. A. Brewer, A.M.Inst.C.E., M.I.M.E., M.I.A.E. 250 pages. With numerous illustrations. Demy 8vo, cloth *Net* **5/0**

MOTOR CAR CATECHISM. Containing about 320 Questions and Answers Explaining the Construction and Working of a Modern Motor Car. For the Use of Owners, Drivers, and Students. By John Henry Knight. Second Edition, revised and enlarged, with an additional chapter on Motor Cycles. Crown 8vo, with Illustrations *Net* **1/6**

The Petrol Engine — Transmission and the Chassis — Tyres — Duties of a Car Driver — Motor Cycles — Laws and Regulations.

MOTOR CARS FOR COMMON ROADS. By A. J. Wallis-Tayler, A.M.Inst.C.E. 212 pp., with 76 Illustrations. Crown 8vo. **4/6**

MOTOR VEHICLES FOR BUSINESS PURPOSES. A Practical Handbook for those interested in the Transport of Passengers and Goods. By A. J. Wallis-Tayler, A.M.Inst.C.E. With 134 Illustrations. Demy 8vo, cloth *Net* **9/0**

Resistance to Traction on Common Roads — Power Required for Motor Vehicles — Light Passenger Vehicles — Heavy Passenger Vehicles — Light Goods Vans — Heavy Freight Vehicles — Self-Propelled Vehicles for Municipal Purposes — Miscellaneous Types of Motor Vehicles — Cost of Running and Maintenance.

OILS AND ALLIED SUBSTANCES. AN ANALYSIS. By A. C. Wright, M.A.Oxon., B.Sc.Lond., formerly Assistant Lecturer in Chemistry at the Yorkshire College, Leeds, and Lecturer in Chemistry at the Hull Technical School. Demy 8vo, cloth *Net* **9/0**

The Occurrence and Composition of Oils, Fats and Waxes — The Physical Properties of Oils, Fats, and Waxes, and their Determination — The Chemical Properties of Oils, Fats, and Waxes from the Analytical Standpoint — Detection and Determination of Non-Fatty Constituents — Methods for Estimating the Constituents of Oils and Fats — Description and Properties of the more Important Oils, Fats, and Waxes, with the Methods for their Investigation — Examination of Certain Commercial Products.

ORGAN BUILDING (PRACTICAL). By W. E. Dickson, M.A., Precentor of Ely Cathedral. Second Edition, Crown 8vo **2/6**

PAINTS, MIXED. THEIR CHEMISTRY AND TECHNOLOGY. By Maximilian Toch. With 60 Photomicrographic Plates and other Illustrations *Net* **12/6**

The Pigments — Yellow, Blue, and Green Pigments — The Inert Fillers and Extenders — Paint Vehicles — Special Paints — Analytical — Appendix.

PAINTING FOR THE IMITATION OF WOODS AND MARBLES. As Taught and Practised by A. R. Van der Burg and P. Van der Burg, Directors of the Rotterdam Painting Institution. Royal folio, cloth, 18½ by 12½ in. Illustrated with 24 full-size Coloured Plates; also 12 Plain Plates, comprising 154 Figures. Fifth Edition *Net* **25/0**

PAINTING, GRAINING, MARBLING, AND SIGN WRITING. With a Course of Elementary Drawing and a Collection of Useful Receipts. By E. A. Davidson. Ninth Edition. Coloured Plates. Crown 8vo, cloth, 5/0; cloth boards, 6/0

PAPER-MAKING. A Practical Manual for Paper Makers and Owners and Managers of Paper-Mills. With Tables, Calculations, etc. By G. Clapperton, Paper-Maker. With Illustrations of Fibres from Micro-Photographs. Second edition, revised and enlarged. Crown 8vo, cloth *Net* **5/0**

Chemical and Physical Characteristics of Various Fibres — Cutting and Boiling of Rags — Jute Boiling and Bleaching — Wet Picking — Washing, Breaking, and Bleaching — Electrolytic Bleaching — Antichlor — Cellulose from Wood — Mechanical Wood Pulp — Esparto and Straw — Beating — Loading — Starching — Colouring Matter — Resin, Size, and Sizing — The Fourdrinier Machine and its Management — Animal Sizing — Drying — Glazing and Burnishing — Cutting, Finishing — Microscopical Examination of Paper — Tests for Ingredients of Paper — Recovery of Soda — Testing of Chemicals — Testing Water for Impurities.

"The author caters for the requirements of responsible mill hands, apprentices, etc., whilst his manual will be found of great service to students of technology, as well as to veteran paper-makers and mill-owners. The illustrations form an excellent feature." — *The World's Paper Trade.*

PAPER-MAKING. A Practical Handbook of the Manufacture of Paper from Rags, Esparto, Straw, and other Fibrous Materials. Including the Manufacture of Pulp from Wood Fibre, with a Description of the Machinery and Appliances used. To which are added Details of Processes for Recovering Soda from Waste Liquors. By A. Watt. With Illustrations. Crown 8vo 7/6

PAPER MAKING, CHAPTERS ON. A Series of Volumes dealing in a practical manner with all the leading questions in connection with the Chemistry of Paper-Making and the Manufacture of Paper. By Clayton Beadle, Lecturer on Paper-Making before the Society of Arts, 1898 and 1902, and at the Battersea Polytechnic Institute, 1902, etc., etc. Each volume is published separately, at the price of 5/0 *net* per vol.

Volume I, comprises a Series of Lectures delivered on behalf of the Battersea Polytechnic Institute in 1902. Crown 8vo. 151 pp. *Net* **5/0**

Volume II. comprises Answers to Questions on Paper-Making Set by the Examiners to the City and Guilds of London Institute, 1901-1903. Crown 8vo, 182 pp. *Net* **5/0**

Volume III. comprises a short practical Treatise in which Boiling, Bleaching, Loading, Colouring, and similar Questions are discussed. Crown 8vo, 142 pp. *Net* **5/0**

CONTENTS: — "Brass" and "Steel" Beater Bars — The Size and Speed of Beater Rolls — The Fading of Prussian Blue Papers — The Effect of Lowering the Breast Roll — The Effect of "Loading" on the Transparency of Paper — "Terra Alba" as a Loading for Paper — The Use of Alum in Tub Sizing — The Influence of Temperature on Bleaching — The Use of Refining Engines — Agitation as an Auxiliary to Bleaching — The Heating of "Stuff" for the Paper Machine — The Comparative Results of Quadruple and Open Effect Evaporation — How to Prevent Electrification of Paper on the Machine — Transparency of Papers — The "Life" of Machine Wires — Edge Runners.

Volume IV. contains discussions upon Water Supplies and the Management of the Paper Machine and its influence upon the Qualities of Papers. Crown 8vo, 164 pp. *Net* **5/0**

CONTENTS: — The Bulking of Papers — Special Qualities of "Art" Papers — The "Ageing" and Storage of Papers — The Use of Lime in Boiling — Controlling the Mark of The "Dandy" — "Machine" and "Hand" Cut Rags — Froth on Paper Machine — Scum Spots in Paper — Consumption of Water in the Manufacture of Paper — The Management of Suction-boxes — The Shrinkage of Paper on the Machine — Paper that does not Shrink or Expand — The Production of Non-Stretchable Paper — The Connection between "Stretch" and "Expansion" of Papers — "Stretch" and "Breaking Strain" — Paper Testing Machines.

Volume V. concerning The Theory and Practice of Beating. Crown 8vo. With photomicrographs and other Illustrations. *Net* **5/0**

CONTENTS: — Early Beating Appliances — The Hollander — The Economy of Beating — Difficulties of arriving at Definite Results — Behaviour of different Fibres — "Refining" — Power Consumption — A Comparison of Two different kinds of Beaters — Power consumed in the "Breaking," "Beating," and "Refining" of different Materials — Dealing with the "Circulation" and "Agitation" in a Hollander — Comparisons of large and medium-sized Hollanders when beating "Hard" and "Soft" Stock — Trials to determine the Relative Merits of Stone and Metal Beater-Bars — Trials with Breakers, Reed Beaters, and Kingsland Refiners — A System of Beating combined with a System for Continuous Bleaching — Beaters and Refiners — Power consumed in grinding Wood-pulp — The Reduction in Length of Fibres at different Stages of Beating — Method for determining the "Wetness" of Beaten Stuff — The Position of Beaters in Old and Modern Paper-Mills — Appendix.

PARA RUBBER. ITS CULTIVATION & PREPARATION. By W. H. Johnson, F.L.S., Ex-Director of Agriculture, Gold Coast Colony, West Africa, Director of Agriculture, Mozambique Company, East Africa, Commissioned by Government in 1902 to visit Ceylon to Study the Methods employed there in the Cultivation and Preparation of Para Rubber and other Agricultural Staples for Market, with a view to Introduce them into West Africa. Second Edition, rewritten and greatly enlarged, with numerous illustrations. Demy 8vo, cloth *Net* **7/6**

The World's Production and Consumption of Rubber — The Para Rubber Tree at Home and Abroad — Propagation — Planting and Cultivating — Soils and Manures — Pests — Latex — Collecting the Latex — Rubber Manufacture — The Antisepticisation of Rubber — Drying and Packing Rubber for Export — Yield of Para Rubber from Cultivated Trees — Establishment and Maintenance of a Para Rubber Plantation — Commercial Value of the Oil in Hevea Seeds.

PASTRYCOOK AND CONFECTIONER'S GUIDE. For Hotels, Restaurants, and the Trade in general, adapted also for Family Use. By R. Wells, Author of "The Bread and Biscuit Baker" 1/0

PETROL AIR GAS. A Practical Handbook on the Installation and Working of Air Gas Lighting Systems for Country Houses. By Henry O'Connor, F.R.S.E., A.M. Inst. C.E., &c., author of "The Gas Engineer's Pocket Book." 80 pages with illustrations. Crown 8vo, cloth *Net* **1/6**

Description of Previous Plants and Systems for Country-House Lighting, Difficulties with, Objections and Prices — History of Petrol Gas, Comparative Costs — Petrol, its Nature, Dangers, and Storing, Notes on the Law regarding same — Burners, Description of same, Piping, Mantles — General Principles of Parts of Plants — Motive Power Meters — Weight-Driven Plants — Root's Blowers — Hot-Air Engines — Pelton Water-Wheels — Descriptions of Various Plants — Extract from an Act for the Safe-Keeping of Petroleum and Other Substances of a Like Nature — Appendix — Useful Notes.

PETROLEUM. THE OIL FIELDS OF RUSSIA AND THE RUSSIAN PETROLEUM INDUSTRY. A Practical Handbook on the Exploration, Exploitation, and Management of Russian Oil Properties, the Origin of Petroleum in Russia, the Theory and Practice of Liquid Fuel. By A. B. Thompson, A.M.I.M.E., F.G.S. 415 pp., with numerous Illustrations and Photographic Plates. Second Edition Revised. Super-royal 8vo, cloth *Net* **21/0**

PETROLEUM MINING AND OIL-FIELD DEVELOPMENT. A Guide to the Exploration of Petroleum Lands, and a Study of the Engineering Problems connected with the Winning of Petroleum. Including Statistical Data of important Oil Fields. Notes on the Origin and Distribution of Petroleum, and a description of the Methods of Utilizing Oil and Gas Fuels. By A. Beeby Thompson,

A.M.I.Mech.E., F.G.S. Author of "The Oil Fields of Russia." 384 pages, 114 illustrations, including 22 full-page plates. Demy 8vo, cloth. [*Just Published.*] *Net* **15/0**

"It is an admirable text-book by a competent authority on an interesting subject." —*Mining Magazine.*

"The present effort is likely to receive a warm welcome in engineering circles, and it can be cordially commended for perusal. It will doubtless have that large sale to which its merits entitle it." — *Mining World.*

"The general aspects of the Petroleum Industry are fully and ably laid out." —*Engineer.*

PIGMENTS, ARTISTS' MANUAL OF. Showing their Composition, Conditions of Permanency, Non-Permanency, and Adulterations, etc., with Tests of Purity. By H. C. Standage. Third Edition. Crown 8vo, cloth 2/6

PORTLAND CEMENT, THE MODERN MANUFACTURE OF. By Percy C. H. West, Chemical Engineer and Consulting Chemist. In Three Volumes. Vol. I., dealing with "Machinery and Kilns." About 200 pages, Medium 8vo. With numerous Illustrations. [*Nearly ready, price about*] **10/6**

PRODUCER GAS PRACTICE (AMERICAN) AND INDUSTRIAL GAS ENGINEERING. By Nisbet Latta, M.Amer.Soc.M.E., M.Amer.Gas Inst. 558 pages, with 247 illustrations. Demy 4to, cloth [*Just Published.*] *Net* **25/-**

Producer Operation—Cleaning the Gas—Works Details—Producer Types—Moving Gases—Solid Fuels—Physical Properties of Gases—Chemical Properties of Gases—Gas Analysis—Gas Power—Gas Engines—Industrial Gas Applications—Furnaces and Kilns—Burning Lime and Cement—Pre-Heating Air—Doherty Combustion Economiser—Combustion in Furnaces—Heat: Temperature, Radiation and Conduction—Heat Measurements: Pyrometry and Calorimetry—Pipes, Flues, and Chimneys—Materials: Fire Clay, Masonry, Weights and Rope—Useful Tables—Oil Fuel Producer Gas.

RECIPES, FORMULAS AND PROCESSES, TWENTIETH CENTURY BOOK OF. Edited by Gardner D. Hiscox, M.E. Nearly

10,000 Scientific, Chemical, Technical, and Household Recipes, Formulas and Processes for Use in the Laboratory and the Office, the Workshop and the Home. Medium 8vo, 800 pp., cloth. *Net* **12/6**

Selected List of Contents: — Absinthe — Acid Proofing — Adhesives — Alcohol — Alkali — Alloys — Aluminium — Ammonia — Aniline — Antidotes for Poison — Anchovy Preparations — Antiseptics — Antiques — Baking powders — Barometers — Beverages — Bleaching — Brass — Brick — Carbolic Acids — Casting — Celluloid — Cheese — Ceramics — Cigars — Coffee — Condiments — Copper — Cosmetics — Cotton — Diamond Tests — Donarite — Dyes — Electro Plating — Embalming — Enamelling — Engraving — Essences — Explosives — Fertilisers — Filters — Food Adulterants — Gelatine — Glass — Gold — Gums — Harness Dressings — Horn — Inks — Insecticides — Iron — Ivory — Jewellers' Formulas — Lacquers — Laundry Preparations — Leather — Linoleum — Lubricants — Matches — Metals — Music Boxes — Oils — Paints — Paper — Perfumes — Petroleum — Photography — Plaster — Plating — Polishes — Porcelain — Poultry — Putty — Rat Poisons — Refrigeration — Ropes — Rubber — Rust Preventives — Salt — Screws — Silk — Silver — Soaps — Solders — Spirit — Sponges — Steel — Stone — Thermometers — Tin — Valves — Varnishes — Veterinary Formulas — Watchmakers' Formulas — Waterproofing — Wax — Weights and Measures — Whitewash — Wood — Yeast.

RUBBER HAND STAMPS. And the Manipulation of Rubber. A Practical Treatise on the Manufacture of Indiarubber Hand Stamps, Small Articles of Indiarubber, The Hektograph, Special Inks, Cements, and Allied Subjects. By T. O'Conor Sloane A.M., Ph.D. With numerous Illustrations. Square 8vo, cloth 5/0

SAVOURIES AND SWEETS. Suitable for Luncheons and Dinners. By Miss M. L. Allen (Mrs. A. Macaire), Author of "Breakfast Dishes," etc. Thirty-first Edition. F'cap 8vo, sewed 1/0
Or, quarter bound, fancy boards **1/6**

SHEET METAL-WORKER'S GUIDE. A Practical Handbook for Tinsmiths, Coppersmiths, Zincworkers, &c., with 46 Diagrams and Working Patterns. By W. J. E. Crane. Crown 8vo, Cloth 1/6

SHEET METAL-WORKER'S INSTRUCTOR. Comprising Geometrical Problems and Practical Rules for Describing the Various Patterns Required by Zinc, Sheet-Iron, Copper, and Tin-Plate Workers. By R. H. Warn. Third Edition. Revised and Further Enlarged by J. G. Horner, A.M.I.M.E. Crown 8vo, 280 pp., with 465 Illustrations, cloth 7/6

SILVERSMITH'S HANDBOOK. Alloying and Working of Silver, Refining and Melting, Solders, Imitation Alloys, Manipulation, Prevention of Waste, Improving and Finishing the Surface of the Work, etc. By George E. Gee. Fourth Edition Revised, Crown 8vo, cloth 3/0

SOAP-MAKING. A Practical Handbook of the Manufacture of Hard and Soft Soaps, Toilet Soaps, etc. With a Chapter on the Recovery of Glycerine from Waste Leys. By Alexander Watt. Seventh Edition, including an Appendix on Modern Candlemaking. Crown 8vo, cloth 7/6

"The work will prove very useful, not merely to the technological student, but to the soap boiler who wishes to understand the theory of his art." —*Chemical News.*

SOAPS, CANDLES, and GLYCERINE. A Practical Manual of Modern Methods of Utilisation of Fats and Oils in the Manufacture of Soap and Candles, and of the recovery of Glycerine. By L. L. Lamborn, Massachusetts Institute of Technology, M.Am.C.S. Medium 8vo, cloth. Fully Illustrated. 706 pages *Net* **30/0**

The Soap Industry — Raw Materials — Bleaching and Purification of Soap-stock — The Chemical Characteristics of Soap-stock and their Behaviour towards Saponifying Agents — Mechanical Equivalent of the Soap Factory — Cold Process and Semi-boiled Soap — Grained Soap — Settled Rosin Soap — Milled Soap-base — Floating Soap — Shaving Soap — Medicated Soap — Essential Oils and Soap Perfumery — Milled Soap — Candles — Glycerine — Examination of Raw Materials and Factory Products.

SOLUBILITIES OF INORGANIC AND ORGANIC SUBSTANCES. A Hand-book of the most Reliable Quantitative Solubility Determinations. Recalculated and Compiled by Atherton Seidell,

Ph.D., Chemist, Hygienic Laboratory, U. S. Public Health Service, Washington, D C. Medium 8vo, cloth, 377 pages *Net* **12/6**

TEA MACHINERY AND TEA FACTORIES. Describing the Mechanical Appliances required in the Cultivation and Preparation of Tea for the Market. By A. J. Wallis-Tayler, A.M.Inst.C.E. Medium 8vo, 468 pp. With 218 Illustrations *Net* **25/0**

"The subject of tea machinery is now one of the first interest to a large class of people, to whom we strongly commend the volume."
— *Chamber of Commerce Journal*.

WAGES TABLES. At 54, 52, 50, and 48 Hours per Week. Showing the Amounts of Wages from one quarter of an hour to sixty-four hours, in each case at Rates of Wages advancing by One Shilling from 4s. to 55s. per week. By Thos. Carbutt, Accountant. Square crown, 8vo, half-bound 6/0

WATCH REPAIRING, CLEANING, AND ADJUSTING. A Practical Handbook dealing with the Materials and Tools Used, and the Methods of Repairing, Cleaning, Altering, and Adjusting all kinds of English and Foreign Watches, Repeaters, Chronographs, and Marine Chronometers. By F. J. Garrard, Springer and Adjuster of Marine Chronometers and Deck Watches for the Admiralty. Second Edition. Revised. With over 200 Illustrations. Crown 8vo, cloth *Net* **4/6**

WATCHES AND OTHER TIMEKEEPERS, HISTORY OF. By J. F. Kendal, M.B.H. Inst. 1/6 boards; or cloth 2/6

WATCHMAKER'S HANDBOOK. Intended as a Workshop Companion for those engaged in Watchmaking and the Allied Mechanical Arts. Translated from the French of Claudius Saunier, and enlarged by Julien Tripplin, F.R.A.S., and Edward Rigg, M.A., Assayer in the Royal Mint. Fourth Edition. Cr. 8vo, cloth 9/0

"Each part is truly a treatise in itself. The arrangement is good and the language is clear and concise. It is an admirable guide for the young watchmaker." — *Engineering*.

WEIGHT CALCULATOR. Being a Series of Tables upon a New and Comprehensive Plan, exhibiting at one Reference the Exact Value of any Weight from 1 lb. to 15 tons, at 300 Progressive Rates, from 1d. to 168s. per cwt., and containing 186,000 Direct Answers, which, with their Combinations, consisting of a single addition (mostly to be performed at sight), will afford an aggregate of 10,266,000 Answers; the whole being calculated and designed to ensure correctness and promote despatch. By Henry Harben, Accountant. Sixth edition, carefully corrected. Royal 8vo, strongly half bound £1 5s.

"A practical and useful work of reference for men of business generally." — *Ironmonger*.

"Of priceless value to business men." — *Sheffield Independent*.

WOOD ENGRAVING. A Practical and Easy Introduction to the Study of the Art. By W. N. Brown. Crown 8vo, cloth. 1/6

www.ingramcontent.com/pod-product-compliance
Lightning Source LLC
Chambersburg PA
CBHW031442210526
45464CB00005B/2299